Advances in Anatomy, Embryology and Cell Biology publishes critical reviews and state-of-the-art surveys on all aspects of anatomy and of developmental, cellular and molecular biology, with a special emphasis on biomedical and translational topics.

The series publishes volumes in two different formats:

- Contributed volumes, each collecting 5 to 15 focused reviews written by leading experts
- Single-authored or multi-authored monographs, providing a comprehensive overview of their topic of research

231

Advances in Anatomy, Embryology and Cell Biology

More information about this series at http://www.springer.com/series/102

Peter Sutovsky
Editor

Cellular and Molecular Basis of Mitochondrial Inheritance

Mitochondrial Disease and Fitness

Springer

Editor
Peter Sutovsky
Division of Animal Sciences
and Department of Obstetrics,
Gynecology and Women's Health
University of Missouri
Columbia, MO, USA

ISSN 0301-5556 ISSN 2192-7065 (electronic)
Advances in Anatomy, Embryology and Cell Biology
ISBN 978-3-030-04569-2 ISBN 978-3-030-04570-8 (eBook)
https://doi.org/10.1007/978-3-030-04570-8

This Springer imprint is published by the registered company Springer Nature Switzerland AG.
The registered company address is: Gewerbestrasse 11, 6330 Cham, Switzerland

Preface: A Mother's Gift

Mitochondria, the cellular power station organelles, arose from bacterial endosymbionts and, in the course of evolution, became the essential source of the energy storing substrate ATP in both prokaryotic and eukaryotic cells. Contrary to equal maternal and paternal contribution of chromosomes in sexually reproducing organisms, most animal taxa show preference for clonal, uniparental, and overwhelmingly maternal inheritance of mitochondria and mitochondrial DNA (mtDNA). Research into maternal mitochondrial inheritance, referred to as the "mitochondrial Eve paradigm," dates back to a 1940 *Nature* paper that explained it as a dispersion and dilution due to disproportionate number of mitochondria in an oocyte, compared to a spermatozoon (reviewed in Song et al. 2016a; Sutovsky and Song 2017). More recently, studies in rodents, ungulates, and primates revealed the proactive, seek-and-destroy mechanisms that target paternal, sperm-borne mitochondria after fertilization (Sutovsky et al. 1996) with the help of the universal protein recycling machinery, the ubiquitin-proteasome system (UPS) (Sutovsky et al. 1999, 2003). Finally, mitochondrial membrane ubiquitination has been linked to autophagic machinery in invertebrate genetic models *Drosophila* and *C. elegans* (Al Rawi et al. 2011, 2012; Politi et al. 2014), eventually revealing that the linkage of ubiquitination to autophagy also regulates postfertilization sperm mitophagy in mammals (Song et al. 2016b). Besides advancing our knowledge of the developmental and evolutionary aspects of mitochondrial function, a deeper understanding of mechanisms promoting clonal mitochondrial inheritance in humans and other species will help manage human mitochondrial health, fitness, and disease, as well as improve production traits and reproductive performance in agriculturally important domestic animals. This volume addresses the diverse yet interwoven aspects of mitochondrial inheritance and function in animals and humans. Readers will notice that the table of contents loosely follows the phylogenetic tree, starting with nematodes and ending with humans. However, the topics of individual chapters are complementary rather than overlapping, addressing the mitophagy mechanism, mitochondrial genome interactions with environment and nutrition, and the importance of mitochondrial function for health, fitness, and reproduction.

Following early work on the role of UPS in the degradation of sperm mitochondria at fertilization, the studies in *C. elegans* (Al Rawi et al. 2011, 2012; Sato and Sato 2011) elegantly (pun intended) linked UPS, which degrades protein molecules one at a time, to autophagy, the self-engulfment machinery that recognizes ubiquitin-tagged protein aggregates and even whole organelles, including mitochondria, for bulk degradation in the autophagic vacuole. The chapter by Jorge Merlet and coauthors from the laboratory of Vincent Galy (Merlet et al. 2019) uses this animal model, advantageous for its relative ease of genome modification and rapid generational turnover, to decipher the pathways that control the degradation of paternally contributed sperm organelles. Fertilization of *C. elegans* oocyte occurs inside the transparent body, which is advantageous for epifluorescence imaging of molecules involved in the fertilization process. The mature oocyte passes through the spermatheca and an amoeboid spermatozoon cell fuses with the oocyte and delivers, among the structures entering the embryo, the sperm mitochondria surrounded by the Golgi-derived, nematode-specific membranous organelles. Subsequently, these sperm-contributed structures are degraded by autophagy, guided by autophagic marker proteins LGG-1 and LGG-2, homologues of mammalian autophagy receptors GABARAP and LC3, respectively. This process may be facilitated by depolarization and proteasome-dependent disruption of mitochondrial membrane, as well as by mtDNA-specific endonucleases. The existence of integrated but distinct mechanisms that converge during sperm mitophagy is proposed, opening up new lines of investigation using both genetic and proteomic approaches.

One of the most notable exceptions to maternal inheritance rule is the bivalvian mollusk *Mytilus,* in which the doubly uniparental inheritance is controlled by specific genetic and molecular mechanisms guiding postfertilization recognition of paternal mitochondria, reviewed in the second chapter by Eleftherios Zourous and George C. Rodakis (Zouros and Rodakis 2019). A *Mytilus* egg comes either with or without the "masculinizing" factor and the "sperm mitochondria binding" factor, a property of the female that is determined by two corresponding nuclear genes that are always in the on/on or the off/off phase. A fertilized egg without these factors develops into a female and inherits sperm mitochondria while still having predominantly maternal mtDNA. The egg endowed with said factors develops into a male, and the sperm mitochondria become segregated into one embryonic blastomere destined to give rise to sperm progenitor germ cells. In this case, paternal mitochondria in the male embryo escape elimination from the germ line, resulting in spermatozoa free of maternal mtDNA, while the soma of the embryo still inherits predominantly maternal oocyte mitochondria. This complex, intriguing inheritance pattern, further complicated by the possibility of mtDNA recombination in mussels, thus flips the insect, nematode, and mammalian models of mitophagy by eliminating maternal mtDNA from male germ cells during embryo development instead of eliminating paternal mitochondria from fertilized egg after fertilization. Further studies of this model will help answer question about the (co)evolution of genes responsible for sexual reproduction, sex determination, nuclear–mtDNA interactions, and allocation of reproductive efforts between sexes.

Continuing with the biparental inheritance theme is a chapter from the research team of William Ballard rooted in research on genetic model *Drosophila*, in which paternal mtDNA inheritance is observed much more frequently than in mammals and under natural conditions rather than by interspecific cross-hybridization or assisted reproduction (Wolff et al. 2012). Similar to other animal models, UPS and autophagy contribute to the elimination of paternal mitochondrial genome in *Drosophila,* when it does occur (Politi et al. 2014). Parent-specific mitochondrial inheritance variations in the fruit fly family can have a bearing on their metabolism, fitness, reproduction, and even speciation and evolution. In the third chapter, Wen C. Aw and coauthors (Aw et al. 2019) review the metabolic aspects of mitochondrial function and homeostasis and the influence of mtDNA mutation on nuclear–mitochondrial cross talk not only in *Drosophila* but also in a wide range of vertebrate species, including humans. Also discussed is the balance between mitochondrial ATP and reactive oxygen species (ROS) production and the sex-biased trade-off between metabolic efficiency and the cost of gamete production. Authors conclude that exogenous factors, such as diet, may differentially influence the selective costs of mtDNA mutations. Further studies of the factors influencing mitochondrial homeostasis are likely to give valuable insight into evolutionary biology and quantitative genetics as well as nutrigenomics and pharmacogenomics.

A gift can sometimes become a curse, quite literally. In the case of mitochondria, the term *mother's curse* has been coined to refer to maternal inheritance of mitochondria with suboptimal mitochondrial genomes. Besides human health, the integrity and functionality of mitochondrial genome has a significant bearing on production traits, fitness, and fertility in economically important livestock species, affecting both economic and sociological aspects of agriculture worldwide. Uniquely, large mammals can now be propagated by somatic cell nuclear transfer (SCNT/cloning) for the purpose of preservation of rare/endangered species and rare breeds, production of transgenic model animals for biomedical research, and mainly to accelerate the dissemination of economically desirable livestock genomes with superior production traits and produce disease-resistant animal strains (Wells and Prather 2017). In the absence of sperm-specific mitophagy determinants, donor cell mitochondria and mtDNA are invisible to mitophagic machinery present in the oocyte cytoplasm, resulting in cloned embryos with heteroplasmy due to the presence of both donor cell and recipient ooplast mitochondrial genomes. In their chapter, Kanokwan Srirattana and Justin St. John address mitochondrial genome inheritance patterns and mitochondrial dysfunction associated with SCNT, focusing on how they impact the reproductive performance in agriculturally important livestock species (Srirattana and St John 2019). Also discussed are the benefits of donor cell mtDNA depletion and mtDNA supplementation to cloned mammalian embryos, aimed at enhancing the efficiency of in vitro embryo production. Apart from technological benefits, the incorporation of mtDNA supplementation into SCNT protocols offers an intriguing model system for studying mitochondrial inheritance and mtDNA–nuclear interactions.

Altogether, the first four chapters solidify the view that sperm mitophagy pathway is well conserved across vertebrate and invertebrate taxa, sharing common molecular components while also displaying unique features and variations on the mitophagy theme (such as the doubly uniparental inheritance in *Mytilus*) that differentiate it from autophagy observed in somatic cells. Providing a fitting punctum and an exclamation mark to this monograph, the closing chapter by Peter Kramer and Paola Bressan puts mitochondrial health and inheritance in the general context of human fitness, lifestyle, and reproductive health (Kramer and Bressan 2019). Filled with thought-provoking ideas and commentaries, this chapter highlights the influences of mitochondrial health and inheritance on animal fitness and human well-being. It is intriguing how the mitochondria-produced ROS delicately balance homeostasis with cellular damage, how prolonged life span may come at the expense of reproduction (and vice versa), and how mitochondria control immune response, food intake, and circadian clocks. In authors' own words, the tiny but mighty mitochondria are so important that an average human carries 14,000 square meters of inner mitochondrial membranes, able to generate membrane potential on a par with the electric field of a lightning bolt. Such a realization sets the stage for discussing the influence of mitochondrial health and integrity on growth, aging, reproduction, health and disease, exercise, sleep, diet, and food restriction. The authors also point out that the ever-so-popular, mitochondria-targeting nutritional supplements and antioxidants are in fact a double-edged sword. Space is given to the discussion of mitochondrial involvement in the reproductive process and strategy, as well as to energy cost of procreation, and to the discussion of assisted reproductive therapies aimed at enabling conception with less than perfect spermatozoa and eggs rejuvenated by mitochondrial replacement therapy. Among many lessons learned, the one close to all of us is that we can live longer, healthier lives simply by offering only the necessary amount of particular substrates (such as glucose) to our own mitochondria, by limiting our food intake and our natural urge to store energy in fat cells. Helpfully, some of the past and current diet fads are discussed in this context. In a captivating and approachable way, the closing chapter provides a compelling manifesto for the importance of mitochondrial research for our health, well-being, and prosperity.

In closing, I would like to thank all contributors and their teams, my fellow *mitochondriacs*, for tackling the task of compiling their research with great enthusiasm and wit. As an editor, I am very grateful for their insightful contributions and, with their busy schedules, their willingness to sacrifice their time and mitochondrion-produced energy to write chapters for this volume. Lastly, I would like to dedicate this monograph to the memory of my mom, Anna, whom I lost not long ago and who gave me the best gifts only mothers can give: life, unconditional love, and *healthy mitochondria*.

Columbia, MO, USA Peter Sutovsky

References

Al Rawi S, Louvet-Vallee S, Djeddi A, Sachse M, Culetto E, Hajjar C, Boyd L, Legouis R, Galy V (2011) Postfertilization autophagy of sperm organelles prevents paternal mitochondrial DNA transmission. Science 334:1144–1147

Al Rawi S, Louvet-Vallee S, Djeddi A, Sachse M, Culetto E, Hajjar C, Boyd L, Legouis R, Galy V (2012) Allophagy: a macroautophagic process degrading spermatozoid-inherited organelles. Autophagy 8:421–423

Aw WC, Garvin MR, Ballard JWO (2019) Exogenous factors may differentially influence the selective costs of mtDNA mutations. Adv Anat Embryol Cell Biol 231:51–74

Kramer P, Bressan P (2019) Mitochondria inspire a lifestyle. Adv Anat Embryol Cell Biol 231:105–126

Merlet J, Rubio-Pena K, Al Rawi S, Galy V (2019) Autophagosomal sperm organelle clearance and mtDNA inheritance in *C. elegans*. Adv Anat Embryol Cell Biol 231:1–24

Politi Y, Gal L, Kalifa Y, Ravid L, Elazar Z, Arama E (2014) Paternal mitochondrial destruction after fertilization is mediated by a common endocytic and autophagic pathway in *Drosophila*. Dev Cell 29:305–320

Sato M, Sato K (2011) Degradation of paternal mitochondria by fertilization-triggered autophagy in *C. elegans* embryos. Science 334:1141–1144

Song WH, Yi YJ, Sutovsky M, Meyers S, Sutovsky P (2016a) The ART and science of sperm mitophagy. Autophagy 12:2510–2511

Song WH, Yi YJ, Sutovsky M, Meyers S, Sutovsky P (2016b) Autophagy and ubiquitin-proteasome system contribute to sperm mitophagy after mammalian fertilization. Proc Natl Acad Sci U S A 113:E5261–E5270

Srirattana K, St John JC (2019) Transmission of dysfunctional mitochondrial DNA and its implications for mammalian reproduction. Adv Anat Embryol Cell Biol 231:75–104

Sutovsky P, Song WH (2017) Post-fertilisation sperm mitophagy: the tale of mitochondrial Eve and Steve. Reprod Fertil Dev 30:56–63

Sutovsky P, Navara CS, Schatten G (1996) Fate of the sperm mitochondria, and the incorporation, conversion, and disassembly of the sperm tail structures during bovine fertilization. Biol Reprod 55:1195–1205

Sutovsky P, Moreno RD, Ramalho-Santos J, Dominko T, Simerly C, Schatten G (1999) Ubiquitin tag for sperm mitochondria. Nature 402:371–372

Sutovsky P, McCauley TC, Sutovsky M, Day BN (2003) Early degradation of paternal mitochondria in domestic pig (Sus scrofa) is prevented by selective proteasomal inhibitors lactacystin and MG132. Biol Reprod 68:1793–1800

Wells KD, Prather RS (2017) Genome-editing technologies to improve research, reproduction, and production in pigs. Mol Reprod Dev 84:1012–1017

Wolff JN, Nafisinia M, Sutovsky P, Ballard JW (2012) Paternal transmission of mitochondrial DNA as an integral part of mitochondrial inheritance in metapopulations of *Drosophila* simulans. Heredity (Edinb) 110:57–62

Zouros E, Rodakis GC (2019) Doubly uniparental inheritance of mtDNA: an unappreciated defiance of a general rule. Adv Anat Embryol Cell Biol 231:25–50

Contents

Autophagosomal Sperm Organelle Clearance and mtDNA Inheritance in *C. elegans*

Jorge Merlet, Karinna Rubio-Peña, Sara Al Rawi, and Vincent Galy

Contents

Abstract The nematode *C. elegans* represents a powerful experimental system with key properties and advantages to study the mechanisms underlying mitochondrial DNA maternal inheritance and paternal components sorting. First, the transmission is uniparental and maternal as in many animal species; second, at fertilization sperm cells contain both mitochondria and mtDNA; and third, the worm allows powerful genetics and cell biology approaches to characterize the mechanisms underlying the uniparental and maternal transmission of mtDNA. Fertilization of *C. elegans* oocyte occurs inside the transparent body when the mature oocyte resumes meiosis I and passes through the spermatheca. One amoeboid sperm cell fuses with the oocyte and delivers its whole content. Among the structures entering the embryo, the sperm mitochondria and a fraction of the nematode-specific membranous organelles are rapidly degraded, whereas others like centrioles and sperm genomic DNA are transmitted. In this chapter, we will review the knowledge acquired on sperm inherited organelles clearance during the recent years using *C. elegans*.

J. Merlet, K. Rubio-Peña, S. Al Rawi, and V. Galy (✉)
Sorbonne Université, CNRS, Institut de Biologie Paris Seine, IBPS, Developmental Biology Laboratory, UMR7622, Paris, France
e-mail: vincent.galy@upmc.fr

P. Sutovsky (ed.), *Cellular and Molecular Basis of Mitochondrial Inheritance*, Advances in Anatomy, Embryology and Cell Biology 231, https://doi.org/10.1007/102_2018_1, Published online: 23 November 2018

Keywords ALLO-1, Allophagy, Autophagy, *C. elegans*, LGG-1, LGG-2, Membranous organelles, Mitophagy, mtDNA inheritance, Sperm mitochondria

1 *C. elegans* mtDNA Inheritance and Fertilization

1.1 C. elegans *to Study Maternal Transmission of mtDNA*

C. elegans mitochondrial DNA (mtDNA) is similar in size and gene content to vertebrates' mtDNA including human (Anderson et al. 1981). It is a 13.8 kb circular molecule carrying 36 genes, including 2 ribosomal RNAs, 22 transfer RNAs, and 12 mitochondrial proteins genes (Okimoto et al. 1992). *C. elegans* germ cells proliferate and start to differentiate in a syncytium. During oocyte maturation, cellularization occurs in the proximal part of the gonadal arm of the hermaphrodite worm (Fig. 1). This cellularization process allows mitochondria incorporation, and around 25,000 copies of the maternal mtDNA are transmitted to the embryo via the oocytes' mitochondria. This copy number remains stable during embryonic and larval development up to the third larval stage (Tsang and Lemire 2002a). The first mtDNA amplification, corresponding to a fivefold increase, occurs at the L4 stage, at the time of gonad activation. A second sixfold amplification occurs in the hermaphrodite adults. This mtDNA amplification is necessary for normal

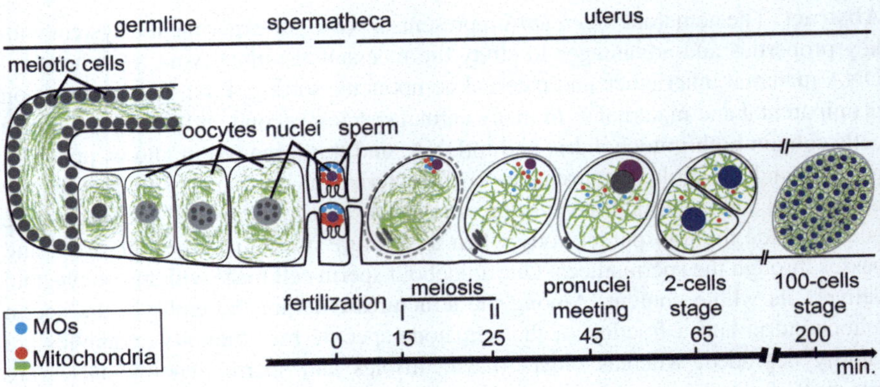

Fig. 1 *C. elegans* to study the fate of sperm components after fertilization. Schematic representation of one of the two gonad arms of a *C. elegans* hermaphrodite adult. The U-shape syncytial gonad produces cellularized oocytes arrested in prophase of the first meiotic division. These oocytes receive maternal mitochondria (green). The most proximal mature oocyte resumes meiosis and is pushed through the spermatheca for fertilization and toward the uterus where it starts embryonic development. Upon fertilization and gametes fusion, MOs (blue), sperm mitochondria (red) and sperm nuclear DNA (violet) enter the oocyte and define the posterior pole of the embryo. Upon meiosis I and II completion, the eggshell is formed, and the sperm-derived organelles are dispersed within the cellular volume and randomly segregated between the two blastomers. By the 100-cell stage (200 min), sperm organelles are degraded

development since inhibition of mtDNA replication by an ethidium bromide treatment induces an L3 arrest (Tsang and Lemire 2002a).

The copy number of mtDNA is sex dependent. Indeed male worms have half of the hermaphrodite's mtDNA (Tsang and Lemire 2002a). This difference is even more spectacular in gametes. The number of mtDNA molecules in *C. elegans* spermatozoa was estimated to be around 30–40 (Tsang and Lemire 2003) compared to the 25,000 copy present in the mature oocyte. Despite their low number, the mtDNA molecules are visible by fluorescent light microscopy as few bright spots in the mitochondria of the mature motile spermatozoa cells using SYBR Green, a DNA intercalant (Fig. 2). At the time of fertilization, sperm mitochondria enter the embryo

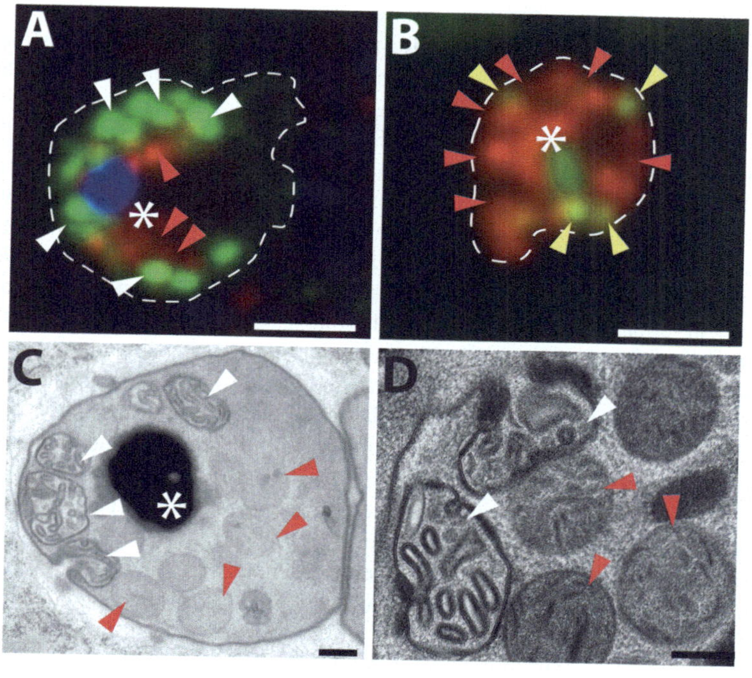

Fig. 2 *C. elegans* spermatozoa contain mitochondria with mtDNA as well as membranous organelles (MOs). Single spermatozoa observed by fluorescent light microscopy (**a** and **b**) or transmission electron microscopy (**c** and **d**). (**a**) MOs (white arrowheads) were visualized using the SP56 antibody (green) on a fixed spermatozoon. Sperm mitochondria (red arrowheads) were labeled using CMXRos (red). Sperm nuclear DNA (asterisk, blue) was labeled with Hoechst. (**b**) Sperm mitochondria (red arrowheads) were labeled using tetramethylrhodamine ethyl ester dye (TMRE, red) on living spermatozoa. Sperm nuclear DNA (asterisk, green) and sperm mitochondrial DNA (yellow arrowheads) were labeled with SYBR Green (green). (**c**) Transmission electron micrograph of a mature spermatozoon after high-pressure freezing and freeze substitution showed membranous organelles fused with the plasma membrane (white arrowheads) and the sperm mitochondria (red arrowheads) around the condensed nuclear chromatin (asterisk). (**d**) At higher magnification the cristae were visible in the sperm mitochondria. Scale bars are 2.5 μm (**a** and **b**), 500 nm (**c**) and 250 nm (**d**)

(Al Rawi et al. 2011; Sato and Sato 2011) along with their mtDNA (Zhou et al. 2011, 2016). Despite the 600- to 800-fold dilution of sperm mtDNA compared to the maternal mtDNA, dilution is not the only mechanism insuring uniparental maternal transmission of mtDNA (Sect. 2).

The isolation of a mtDNA mutant strain was instrumental to track sperm mtDNA in the progeny. The most frequently used strain contains the *UaDf5* mtDNA allele which carries a 3 kb deletion affecting 4 proteins-encoding genes and 11 tRNA-coding genes. Interestingly, this shorter molecule is maintained at a high copy number in the worms together with wild-type (WT) mtDNA molecules which complement the not fully functional *UaDf5* allele. Therefore, animals carrying *UaDf5* allele are heteroplasmic, with the *UaDf5* representing 60% of the total number of mtDNA molecules which is twice of the total mtDNA molecules in WT worms (Tsang and Lemire 2002b). Because of its huge size, the *UaDf5* deletion can be easily tracked by PCR. Historically, this allele was used to demonstrate that mtDNA is maternally transmitted in worms, and then it rapidly became a tool of choice to study the mechanisms of sperm mtDNA clearance (Al Rawi et al. 2011; Djeddi et al. 2015; Sato and Sato 2011; Sato et al. 2018; Zhou et al. 2011, 2016). When heteroplasmic males carrying the *UaDf5* allele are crossed with WT hermaphrodites, the progeny only contains the WT mtDNA after the 64-cell stage (Zhou et al. 2011). This sperm mtDNA clearance in the embryo to insure strictly maternal inheritance of mtDNA is correlated with the degradation of labeled sperm-derived mitochondria (Al Rawi et al. 2011; Sato and Sato 2011; Zhou et al. 2011; Sect. 2).

C. elegans is an androdioecious (male-hermaphrodite) specie with self-fertilizing hermaphrodites. Neither in self-fertilization nor in cross fertilization, the sperm-derived mtDNA are kept in the progeny, and no differences in the timing and mechanism of sperm mitochondria degradation were detected despite their different origins. In other words, the degradation of the sperm mitochondria and the nematode-specific Golgi-derived membranous organelles (MOs) occurred also when sperm mitochondria had the same origin than the oocyte mitochondria. This characteristic strongly suggests that the mechanisms responsible for sperm mitochondria degradation recognize sperm property/ies rather than their exogenous origin.

The potential reason for uniparental maternal heredity could be the need to avoid the transmission of mutated forms of the mtDNA generated during the life of the motile spermatozoon. The coexistence of different forms of mtDNA molecules, a.k.a. heteroplasmy, is thought to be deleterious and unstable in the organism (Sharpley et al. 2012). Even though, the degree of mitochondrial heteroplasmy in natural worm populations and within single individuals has not been documented, studies of the heteroplasmic worms carrying the *UaDf5* mtDNA allele revealed defects (Liau et al. 2007; Sato et al. 2018). This heteroplasmic strain is viable, shows reduced egg-laying and defecation rates, a shorter life-span, as well as a reduced motility of sperm cells (Liau et al. 2007). Furthermore, the duration of cell division is extended, and the embryonic lethality is higher in these heteroplasmic worms with 9.4% of lethality against 0.4% for homoplasmic wild-type worms (Zhou et al. 2016). However, it is worth noting that this heteroplasmy may represent a very peculiar case since it is stable over generations (Tsang and Lemire 2002b), while heteroplasmy in mice has been described as very unstable with the rapid switch to a homoplasmic status (Sharpley et al. 2012). This stability in the worms carrying the *UaDf5* mtDNA is probably the result of antagonist selection mechanisms. The smaller mtDNA has a replicative advantage, while the WT mtDNA remains required for cell/organism viability (Tsang and Lemire 2002b). Interestingly, milder embryonic phenotypes (embryonic lethality, cell division duration) were also observed for embryos showing heteroplasmy due to the transient stabilization of sperm mtDNA from a different wild-type isolate (Hawaïn crossed with the Bristol isolate) (Zhou et al. 2016). This experiment demonstrates that transient heteroplasmy induced by a delayed degradation of sperm mitochondria is deleterious for the embryo. Testing the impact of heteroplasmy over several generations has not been possible so far since sperm mitochondria are still degraded later during development, and some of the mutations or RNAi delaying their degradation, like autophagy inactivation, are lethal by themselves (Al Rawi et al. 2011; Sato and Sato 2011).

1.2 Sperm Components Entering Embryo at Fertilization

C. elegans spermatozoa are amoeboid cells of around 5 μm. The spermatozoa are formed at the end of the fourth larval stage in hermaphrodites or in males. Four spermatids form the budding of one secondary spermatocyte. During the budding, several cellular components present in the spermatocytes like actin, tubulin, and the ribosomes are excluded from the forming spermatids and retained in the residual body (for review, see Nishimura and L'hernault 2017). The nuclear DNA, the MOs, and the mitochondria are segregated in the four spermatocytes and remain visible in the mature spermatozoon (Fig. 2). The mature spermatozoa correspond to the spermatids activated into mobile spermatozoa at the time of the first ovulation or upon ejaculation of the male with TRY-5 protease in the seminal fluids (Smith and Stanfield 2011).

In hermaphrodite worms, 99% of the spermatozoa are used to fertilize an oocyte (Ward and Carrel 1979). When males and hermaphrodites are mated, the spermatozoa from the male displace the ones from the hermaphrodite (Ward and Carrel 1979). At each fertilization, some of the spermatozoa are pushed out of the spermatheca toward the uterus (Ward and Carrel 1979) and have then to crawl back toward the spermatheca to get ready for the next fertilization event (Ward and Carrel 1979). The motility relies on the polymerization activity of the proteins of the family of the major sperm proteins (MSP) representing up to 40% of the total protein composition of mature sperm cells (Italiano et al. 1996). MSP assembly into fibers depends on ATP, and therefore sperm motility requires mitochondria activity for ATP supply. This production of ATP might generate ROS and represent a threat for sperm mtDNA integrity. At fertilization, the plasma membrane of the spermatozoon fuses with the plasma membrane of the oocyte, and its content enters the cytoplasm. This includes, at least, the MSP proteins, the centrioles, the genomic DNA, the sperm mitochondria with their mtDNA molecules, and the nematode-specific MOs.

MSP proteins, sperm mitochondria, and MOs disappear soon after fertilization, while sperm genomic DNA and the centrioles are critical for embryogenesis. In this chapter, we will focus only in the organelles that are present in the mature spermatozoa and degraded after their entry.

1.2.1 The Membranous Organelles (MOs)

The membranous organelles (MOs) are nematode-specific sperm organelles with a bilobed structure composed of a smaller head that is separated by an electron-dense collar from a larger body (Fig. 2). MOs are Golgi-derived vesicles that fuse with the plasma membrane during the maturation of spermatocytes into mature spermatozoa. During spermatogenesis MOs are associated with the fibrous body (FB) and form the FB-MO complex involved in partitioning the cytoplasm during spermatogenesis. MOs also play a critical role during the asymmetric cell division occurring in sperm meiosis II (Ward et al. 1981; Wolf et al. 1978). One major role of the FB-MO complex is to ensure the proper segregation of proteins into the spermatids rather than to the residual body, and during spermatozoa maturation, new components are added to the cell surface by MO fusion to the plasma membrane (Chatterjee et al. 2005; Roberts et al. 1986; Shakes and Ward 1989; Ward et al. 1981; Xu and Sternberg 2003). Therefore, MO morphology changes dramatically during spermatogenesis, and blocking MO fusion results in non-motile spermatozoa unable to adhere to the uterine walls (Machaca and L'hernault 1997; Washington and Ward 2006).

Even when most MOs are fusing upon sperm maturation, a fraction remains unfused as well as the body of the MOs (Nelson and Ward 1980) and still contains at least the SP56 antigens. So, some MOs enter the ooplasm at fertilization and are specifically targeted for degradation. One could speculate that MO membranes and associated proteins could be harmful for the embryo. For example, MO could fuse with membranous compartments in the oocyte and alter their functions. Indeed, FER-1, which contains six C2 domains (protein kinase C conserved region 2) domains which usually act in calcium-dependent lipid processing events, such as vesicle fusion (Lemmon 2008), is one of the identified proteins localized in MOs (Washington and Ward 2006). FER-1 regulates the calcium-dependent membrane fusion between MOs and the plasma membrane. Therefore, after fertilization and the entry of unfused MO into the cytoplasm of the oocyte, the degradation of the MOs would be a mechanism to get rid of FER-1 and avoid their inappropriate fusion with embryo vesicles.

1.2.2 Sperm Mitochondria

Mitochondria in *C. elegans* spermatozoa are distributed around the condensed chromatin and excluded, like the MOs, from the pseudopod region. TEM analysis revealed that they tend to have circular sections, while the mitochondria of maternal origin in the embryo show sections of elongated shapes. The average diameter of the wild-type sperm mitochondria is around 460 nm which makes them distinguishable from the maternal ones which are more tubular with an average diameter of around 240 nm (Zhou et al. 2016).

The observation of the mitochondria morphology by light microscopy shows clear differences between oocyte and sperm-derived mitochondria within the 1-cell stage embryo completing meiosis division (Al Rawi et al. 2011; Sato and Sato 2011; Fig. 3). The few sperm mitochondria are localized around the condensed sperm chromatin at the posterior side of the embryo (Fig. 3), and they appear more fragmented than the elongated and interconnected maternal mitochondria (Fig. 4). The sperm mitochondria show the presence of clear cristae by TEM and are labeled by fluorescent dyes accumulating into mitochondria with a membrane potential arguing for intact, functional, and polarized sperm mitochondria prior fertilization (Al Rawi et al. 2011; Sato and Sato 2011).

2 Sperm Organelles Clearance

2.1 Autophagy Degradation of MOs and Sperm Mitochondria

2.1.1 Kinetic of Sperm Organelles Clearance

One remarkable property of the process in *C. elegans* compared to other species is its speed. Sperm mitochondria and MOs are rapidly degraded in the embryo before the 64-cell stage in less than 3 h of development (Al Rawi et al. 2011; Sato and Sato 2011). By comparison sperm mitochondria are degraded by 84 h in the mouse embryo (Luo et al. 2013; Rojansky et al. 2016). In *Drosophila*, the sperm mtDNA is eliminated prior to fertilization, and the mitochondria are degraded in the embryo within few hours (Politi et al. 2014).

The LGG-1 and LGG-2 autophagy proteins (*C. elegans* homologs of LC3/Atg8 proteins) are recruited around MOs and sperm mitochondria at the end of the first female meiosis division (15–20 min post fertilization) (Al Rawi et al. 2011; Sato and Sato 2011; Fig. 5). LGG-1 is required for autophagosome formation and LGG-2 recruitment, while LGG-2 is dispensable. Interestingly, these respective contributions in the formation of autophagosomes are correlated with the impact of their depletion on embryo development. LGG-1 is essential, while worms lacking LGG-2 are viable (Djeddi et al. 2015; Manil-Segalen et al. 2014). This suggests that their respective function in sperm-inherited organelle degradation is characteristic of their function in other autophagic processes during development.

In the 1-cell stage embryo, the localization of the autophagy markers is restricted to the area around sperm-inherited organelles. LGG-1 is localized in few cytoplasmic foci in the oocytes (Al Rawi et al. 2011; Sato and Sato 2011). After fertilization LGG-1 and LGG-2 are essentially localized around sperm organelles at the posterior pole of 1-cell embryos ongoing first and second meiotic divisions (Al Rawi et al.

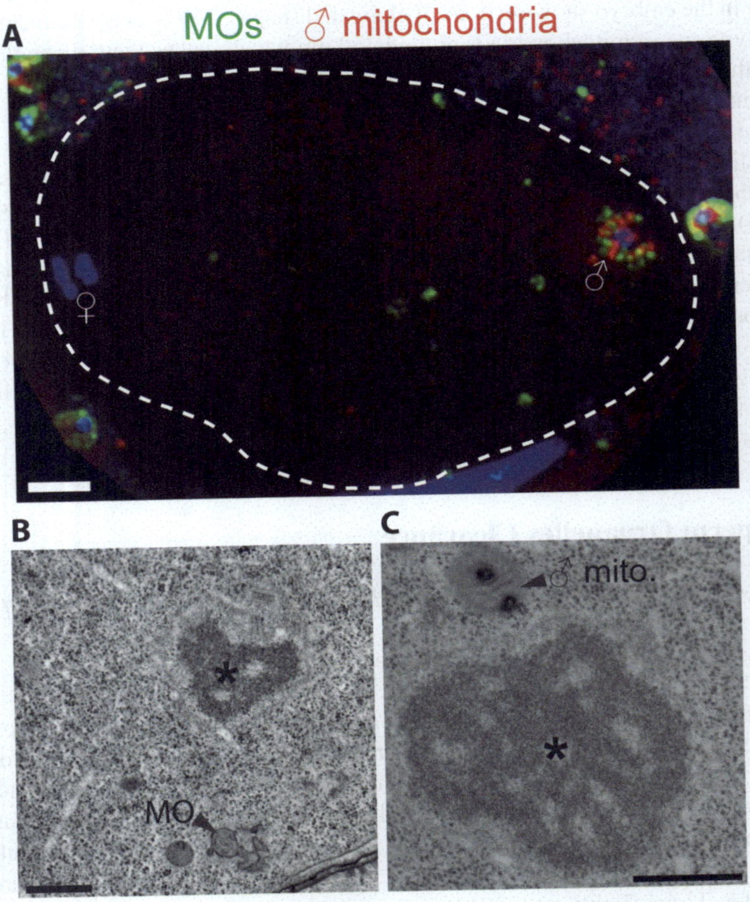

Fig. 3 MOs and sperm mitochondria are found in the cytoplasm, at the posterior pole of the 1-cell stage *C. elegans* embryo shortly after fertilization. (**a**) MOs and the sperm mitochondria were observed in spermatozoa outside the embryo and inside a 1-cell stage embryo by confocal microscopy. MOs were visualized using the SP56 antibody (green) on a fixed sample, sperm mitochondria were labeled using CMXRos (red), and nuclear DNA was labeled with Hoechst (blue). At the end of the second meiotic division (around 30 min after fertilization), MOs and sperm mitochondria are still grouped around the sperm nuclear DNA, at the posterior pole of the 1-cell embryo, ready to be targeted by the autophagy machinery. The dotted line indicates the border of the embryo; sperm- (♂) and oocyte-derived (♀) nuclear DNA are indicated. (**b and c**) Transmission electron micrograph of spermatozoid organelles around sperm nuclear chromatin (asterisk) in a 1-cell embryo at metapase of the first meiosis division, revealing the presence of (**c**) sperm-derived mitochondria (♂mito) with granules in their matrix and (**b**) membranous organelles (MO). Scale bars are 2.5 μm in (**a**) and 500 nm in (**b and c**)

2011; Sato and Sato 2011). Therefore, in *C. elegans*, autophagy degradation of sperm organelles represents the main ongoing autophagy at this time. This is different to the crowded autophagy pattern observed in early developing mouse

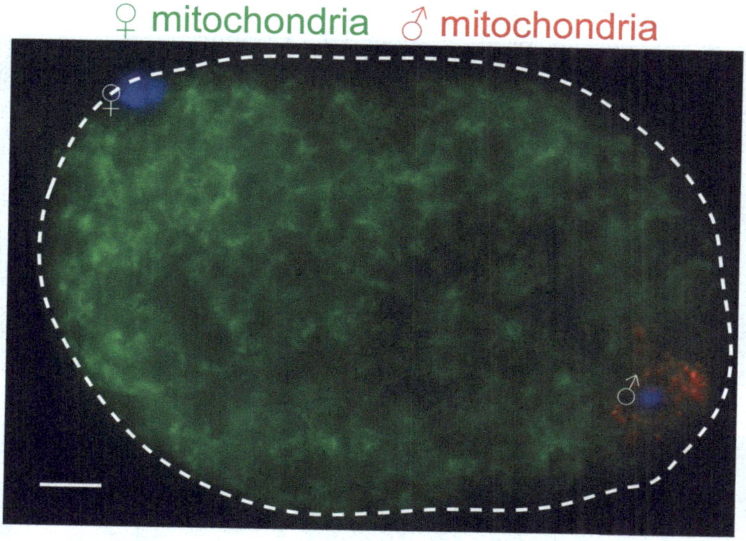

Fig. 4 Following fertilization, *C. elegans* sperm mitochondria are outnumbered by the oocyte-derived mitochondria and show a distinct morphology. Oocyte-derived mitochondria labeled with GFP::ANT-1 protein (green) and sperm-derived mitochondria labeled with CMXRos mitotracker (red) were visualized in a fixed 1-cell stage embryo by fluorescent microscopy. DNA was labeled with Hoechst (blue). Scale bar represents 5 μm

embryos where autophagy has been proposed to degrade maternal proteins (Tsukamoto et al. 2008). Along the same line, inactivation of autophagy (conditional KO of *Atg5*) in mouse embryo prevents the development beyond the 8-cell stage (Tsukamoto et al. 2008). In *C. elegans*, *lgg-1* and *lgg-2* RNAi-treated hermaphrodites produce a progeny showing 80% of late embryonic lethality and 20% of L1 arrest (Al Rawi et al. 2011; Sato and Sato 2011, VG unpublished results). *lgg-1* is a maternal effect gene: the embryonic and larval development are possible due to the maternal contribution from the heterozygote mother. Indeed, *lgg-1* homozygote mutant embryos develop into sterile adults. Therefore, bulk autophagy is, like in vertebrates, an essential process for worm development (Palmisano and Melendez 2018). On the other hand, it was possible to demonstrate that autophagy and the lysosomal pathway are specifically involved in sperm organelle clearance since sperm organelle autophagy is the most prominent ongoing autophagy process in the first 3 h after fertilization. Indeed the inactivation of these pathways using mutants or RNAi depletions allowed the stabilization of sperm mitochondria and their genome as well as of the MOs (Al Rawi et al. 2011; Sato and Sato 2011; Zhou et al. 2011; Fig. 6), while it has no stabilization effect on the degradation of a family of very abundant sperm proteins that enter the embryos at fertilization, the MSP proteins (Al Rawi et al. 2011). The MSP proteins are degraded between the 2- and 4-cell stage by an autophagy-independent process (Al Rawi et al. 2011).

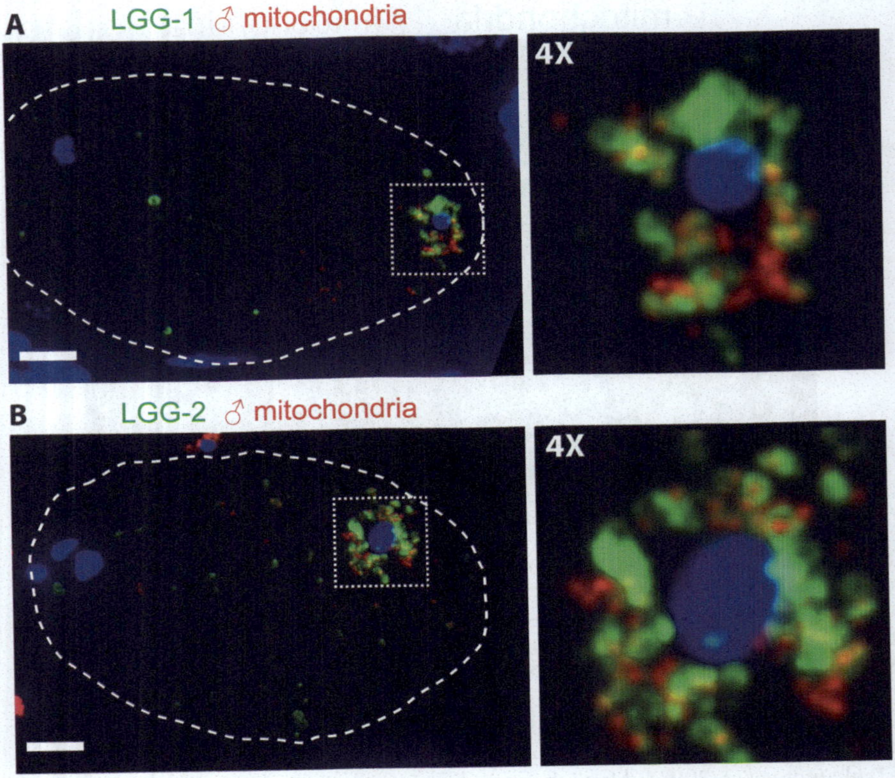

Fig. 5 Autophagy markers are recruited around sperm mitochondria at the posterior pole of *C. elegans* 1-cell stage embryo. LGG-1 (**a**) and LGG-2 (**b**) labeled using specific antibodies (green) in a fixed 1-cell stage embryo were visualized by fluorescent microscopy. Sperm-derived mitochondria were labeled with CMXRos mitotracker (red), and DNA was labeled with Hoechst (blue). Scale bar represents 5 μm

2.1.2 Autophagosome Formation, Maturation, and Dynamics

In WT embryos, the sperm-inherited organelles are initially clustered around the sperm nuclear chromatin at the posterior pole of the embryo (Fig. 3). The autophagy induction depends on the entry of sperm components in the embryo. This was demonstrated using a *spe-9* fertilizing mutant. The *spe-9(hc52ts)* is a thermosensitive mutant that prevents the fusion of the spermatozoa with the oocyte. At restrictive temperature, it accumulates unfertilized oocytes in the uterus, and no GFP::LGG-1 accumulation was observed in these oocytes (Sato and Sato 2011). On the other hand, oocyte fertilization by two spermatozoa induced upon *egg-4/5 (RNAi)* depletion triggers the localization of endogenous LGG-1 protein around the sperm organelles of both spermatozoa (Al Rawi et al. 2011). *egg-4/5(RNAi)* polyspermic embryos also allowed to demonstrate that the autophagy response is independent of

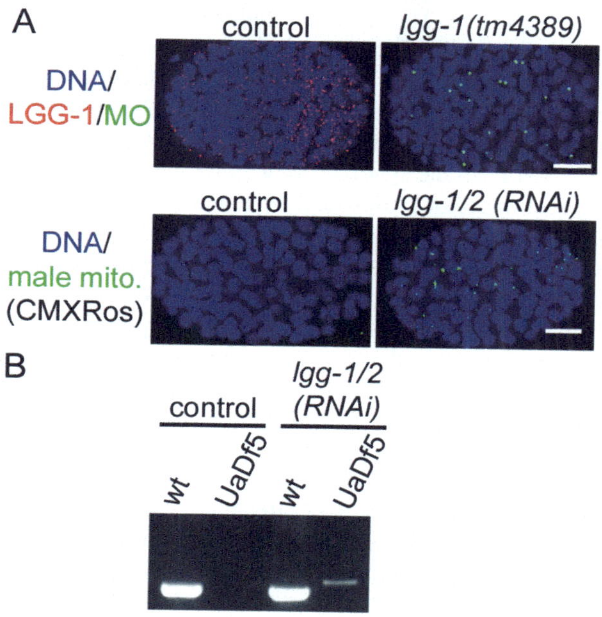

Fig. 6 Autophagy is required to degrade *C. elegans* spermatozoid-inherited organelles after fertilization and prevents mitochondrial heteroplasmy. (**a**) In 120-cell embryos, the MOs (green) are absent in control (left) but remain in *lgg-1(tm3489)* mutant embryo. LGG-1 (red) is present in WT but not in *lgg-1(tm3489)*. Pictures are Z-projections of confocal stacks. Paternal mitochondria (green) from labeled males are absent in control (left) but remain in lgg-*1 + 2(RNAi)* embryo. Scale bar are 10 μm. (**b**) Interfering with autophagy maintains paternal mitochondrial heteroplasmy. Heteroplasmic males carrying both deleted (*UaDf5*) and WT mtDNA were crossed with either control or *lgg-1 + 2(RNAi)* hermaphrodites, and their embryos progeny were tested by PCR for the presence of wild-type (WT) and mutated (*UaDf5*) mtDNA. Male *UaDf5* mtDNA is detected in autophagy-deficient embryos, but not in control, indicating an heteroplasmic state. Adapted with permission from Al Rawi et al. (2011)

the posterior position of the substrates (Al Rawi et al. 2011). The same double induction of autophagy was also observed in polyspermic embryos obtained in the *spe-11(hc77ts)* embryos (Sato and Sato 2011), a protein not required for fertilization but for embryo development. The induction of this early autophagy is also independent of cell cycle progression since, in arrested embryos at metaphase of meiosis I upon *emb-27(RNAi)* treatment, GFP::LGG-1 was still recruited around the sperm organelles (Sato and Sato 2011).

Electron tomographs revealed the modification of sperm mitochondria shortly after their entry in the embryos. The cristae gradually lose their structure, and electron-dense granules appear in the matrix (Zhou et al. 2016; Fig. 3). These types of granules were observed previously (Fawcett 1981) and their origin and potential function discussed (Jacob et al. 1994). They are probably composed of phospholipids and able to bind calcium, but their biological functions are still unknown. These aggregates are growing in size, while the cristae are cleared prior to autophagosome enclosure. This made the authors suggest and test the hypothesis that the paternal elimination begins as a self-destruction process. A dramatic change in morphology and ions exchanges in sperm mitochondria at the time of fertilization has been described in acidians (Lambert and Epel 1979). In these marine species when sperm attach to the chorion, it triggers the swelling of the mitochondrion and its physical exclusion from the spermatozoon along the flagellum. This swelling is coupled with the loss of protons from the mitochondria and sperm cell. This reaction can be triggered by increasing the pH or lowering the Na+ of the sea water. One could speculate that the same type of reaction is triggered on *C. elegans* sperm mitochondria upon their entry.

The autophagy of the sperm-inherited organelles is spontaneous and physiological, making it an attractive model to study macro-autophagy. It has been used to decipher the specific functions of LGG-1 and LGG-2 in the allophagosome formation and maturation. LGG-1 and LGG-2 are both localized around sperm organelles but do not completely overlap (Djeddi et al. 2015). *lgg-2* RNAi depletion reduced the viability of the dauer (a stage of developmental arrest) worms and the lifetime of the adults (Alberti et al. 2010). *lgg-2* RNAi treatment was lethal in mutant carrying a loss-of-function mutation in *daf-2*, the *C. elegans* insulin-like tyrosine kinase receptor, which triggers abnormal constitutive dauer entry (Melendez et al. 2003). LGG-2 function in autophagy was further supported since *lgg-2* RNAi depleted embryos abnormally accumulate P-granules components in the somatic cells (Zhang et al. 2009). The characterization of the *lgg-2(tm5755)−/−* embryos suggested that LGG-2 is required for allophagosomes acidification, a function that would be mediated by an interaction with VPS-39 (Manil-Segalen et al. 2014), one component of the HOPS complex, a multimeric tethering protein complex involved in vesicle fusion of late endosomes. In the absence of LGG-2 protein, LGG-1 was still recruited around the substrates that appeared clustered in the early embryos (Djeddi et al. 2015; Manil-Segalen et al. 2014). This clustering phenotype of the allophagosomes was proposed to be the consequence of a defect in allophagosomes acidification due to the lack of fusion with the lysosomes as a result of the loss of interaction with VPS-39 (Manil-Segalen et al. 2014). The potential link with the endosomal pathway was intriguing and could be conserved among species as suggested by the localization of the VPS-27 in potential hybrid compartments between endosomes and autophagosomes (amphisomes) around sperm-inherited organelles in the 1-cell stage *C. elegans* embryos (Manil-Segalen et al. 2014) as well as the colocalization of Rab7 with sperm mitochondria in *Drosophila* embryos (Politi et al. 2014).

Live embryo imaging of labeled sperm mitochondria and GFP::LGG-1 allowed to describe the dynamics and distribution of the allophagosomes during the first hour of *C. elegans* development. The allophagomes are formed around the clustered sperm organelles near the condensed sperm nuclear DNA at the posterior pole of the embryo. The substrates and the autophagosomes tend to migrate toward the anterior pole of the embryos at the time of pronuclei formation. After pronuclei meeting and during pronuclei centration, the autophagosomes tend to be gathered by the active centrosomes. During the first mitosis, they are randomly distributed between the two daughter blastomers (Djeddi et al. 2015; Hajjar et al. 2014). In the absence of LGG-2, the dynamics of the allophagosomes was dramatically impaired. LGG-1-labeled autophagosomes remained clustered close to the 1-cell stage plasma membrane, and their dispersion between P1 and AB blastomers was abnormal (Djeddi et al. 2015). This defect in the intracellular dynamics was correlated with a delay in the degradation of the sperm-derived MOs and mitochondria (Djeddi et al. 2015). This extended clustering phenotype is also coupled to a block in the retrograde movement of the autophagosomes toward the pericentrosomal area where the acidic compartment tends to accumulate. This suggests that the delay in sperm-derived organelles clearance could be due to a delay in the fusion of autophagosomes with acidic compartments (Djeddi et al. 2015).

2.1.3 Common Recognition Signal of Sperm Organelles

ALLO-1/IKKE-1

The recent identification of IKKE-1, the worm homologue of the TBK1/IKKε protein kinase family, and ALLO-1, an autophagy receptor, provides important information on the mechanism of induction of sperm organelle-specific autophagy (Fig. 7). These factors are expressed in the germ line and required for normal elimination of MOs and sperm mitochondria in early embryogenesis, before 8- or 16-cell stage (Sato et al. 2018). The ALLO-1 protein is maternally contributed and recruited around sperm-inherited organelles within 10 min post fertilization. This is around 5 min before the recruitment of LGG-1 protein, an autophagy marker (Al Rawi et al. 2011; Sato and Sato 2011). The recruitment still occurs even upon the depletion of ATG-11, an early autophagy factor (Sato et al. 2018). ALLO-1 recruitment on the autophagy substrates is therefore a very early step in the process, independent of the autophagy machinery. ALLO-1 contains an N-terminal LC3-interacting region (LIR) responsible for its binding to LGG-1 but dispensable for its localization on the substrates (Sato et al. 2018).

Recognition of ALLO-1 by LGG-1 required ALLO-1 phosphorylation by IKKE-1, but other kinase(s) might be involved. Indeed, a GFP::IKKE-1 kinase dead mutant still localizes to the sperm organelles in the embryo but delays their degradation (Sato et al. 2018). Even if IKKE-1 is not the only kinase responsible for ALLO-1 phosphorylation, it participates in its regulation (Sato et al. 2018).

Ubiquitination of the Sperm Organelles

Prior to fertilization, MOs are ubiquitinated, as shown by immunofluorescence where anti-ubiquitin antibodies decorate MOs in the sperm (Al Rawi et al. 2011; Sato and Sato 2011). After fertilization, a new wave of branched poly-ubiquitin deposition recognized by K48- and K63-specific antibody arises (Al Rawi et al. 2011; Hajjar et al. 2014).

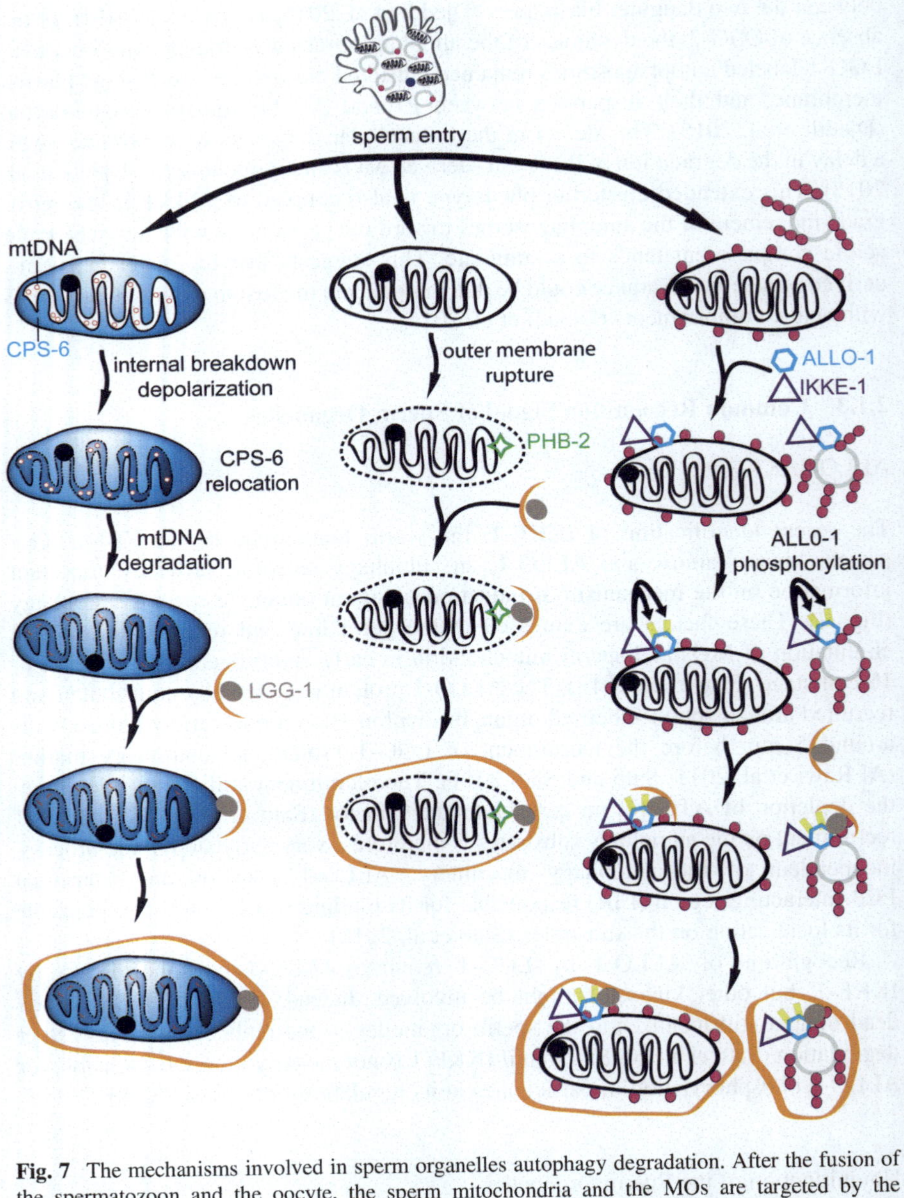

Fig. 7 The mechanisms involved in sperm organelles autophagy degradation. After the fusion of the spermatozoon and the oocyte, the sperm mitochondria and the MOs are targeted by the autophagy machinery with the recruitment of the membrane-associated LGG-1 protein. Several structural modifications and mechanisms were recently described, but more work is required to evaluate their respective contribution and relationship. In all pathways, sperm mitochondria lose their membrane potential (not represented) and accumulate in their matrix electron-dense granules (black dots) of unknown origin and function. (Left) The sperm provided endonuclease G (CSP-6, blue) is relocalized in the matrix after fertilization to degrade mtDNA (red circles). This contributes to sperm mtDNA clearance and sperm mitochondria cristae destabilization before autophagosome

In vertebrates, the ubiquitin mark on sperm mitochondria has been described (Sutovsky et al. 1999). Therefore, the role of the ubiquitin modifications for sperm mitochondria recognition and degradation in *C. elegans* has been considered but then rejected based on the absence of obvious ubiquitin marks on fixed samples (Al Rawi et al. 2011; Hajjar et al. 2014; Sato and Sato 2011), the colocalization of GFP::ubiquitin with membranous organelles, and the absence of clear association with the sperm mitochondria (Al Rawi et al. 2011). However, this has been recently revisited (Sato et al. 2018; Fig. 7). Using GFP-specific antibodies, traces of GFP:: ubiquitin signal were found to be associated with sperm mitochondria after their entry suggesting that sperm mitochondria may contain a low level of ubiquitinated proteins (Sato et al. 2018). In addition, upon RNAi depletion of UBA-1, the unique E1-ubiquitin-conjugating enzyme in worms, the recruitment of autophagy adaptor protein ALLO-1 was prevented (Sato et al. 2018). Ubiquitination was therefore proposed as a signal for ALLO-1-dependent autophagy on MOs as well as sperm mitochondria (Sato et al. 2018). While ubiquitin marks already decorate MOs in activated sperm cells, it is not known whether sperm mitochondria are already marked before fertilization (Sato et al. 2018) and if ubiquitin serves as the first mark for sperm mitochondria targeting. Interestingly proteasome subunits were visualized in spermatozoa and around sperm organelles after their entry (Al Rawi et al. 2011; Hajjar et al. 2014). With two different fluorescent tags allowing to visualize the sperm and the oocyte contributed proteasomal 19S subunits RPT-1, different localizations emerged: the maternally provided RPT-1 localized on MOs while the sperm-contributed fraction around sperm organelles (Hajjar et al. 2014). This dual localization suggests dual function of the proteasome in sperm organelle clearance, but more work is required to identify the potential substrate(s) (Hajjar et al. 2014). These observations were done using RPT-1 subunits expressed under a germ line-specific promotor. It would be interesting to confirm these results using tagged versions of several proteasome subunits expressed under their endogenous promotors using CRISPR-Cas9 genome editing. Interestingly, in Hela cells, LC3 is recruited to the inner mitochondrial membrane via its interaction with the LC3-interacting region (LIR) of the mitochondrial protein PHB2, and this occurs at sites of proteasome-dependent ruptures of the outer mitochondrial membrane (Wei et al. 2017), a process that might depend on ubiquitination. The role of ubiquitination of sperm mitochondria proteins in *C. elegans* has to be confirmed.

Fig. 7 (continued) formation. (Center) Maternal prohibitin 2 (PHB-2), an internal mitochondrial membrane protein, is exposed to the cytosol due to the outer membrane rupture. PHB-2 might participate in LGG-1 recruitment and autophagosome formation. (Right) Sperm mitochondria and MOs are ubiquitinated in the ooplasm and recruit the ALLO-1 autophagy adaptor (hexagon) and its kinase IKKE-1 (circle). IKKE-1 phosphorylates ALLO-1, and this contributes to autophagosome assembly and sperm organelle degradation. ALLO-1 recruits LGG-1 and associated membranes via its LIR domain. CSP-6 and PHB-2 are specific of the sperm-derived mitochondria, while ALLO-1/IKKE-1 are involved in the targeting of both the MOs and the sperm-derived mitochondria

2.2 *Specific Mechanism of Sperm Mitochondria Degradation*

2.2.1 Loss of Mitochondrial Membrane Potential

A common signal in mitophagy is the loss of membrane potential. This can be evaluated using tetramethylrhodamine ethyl ester (TMRE), a fluorescent dye only imported in mitochondria with a membrane potential. Therefore, the hypothesis that sperm mitochondria lose their membrane potential was evaluated. A loss of membrane potential based on the loss of TMRE signal from sperm mitochondria after their entry into the embryo was observed (Sato et al. 2018; Zhou et al. 2016). However, no direct observation of this loss of TMRE staining has been observed in utero. Interestingly, the loss of membrane potential was also observed in mouse embryo between 18 and 48 h of development suggesting that the loss of membrane potential might be a common event in sperm mitochondria degradation. Even if the sequence appeared conserved, the timing is different since the loss of membrane potential seems to occur in less than 10 min in *C. elegans* (Sato et al. 2018), while it takes more than 18 h in mice (Rojansky et al. 2016). It must be noticed that there are yet no direct experimental evidences that the loss of membrane potential has a functional role in sperm mitochondria degradation (Rojansky et al. 2016).

When males were labeled using TMRE and SYTO11 and crossed with unlabeled hermaphrodites, no TMRE signal was colocalized with the SYTO11 signal in the 1-cell stage embryo (Zhou et al. 2016) suggesting the absence of TMRE signal in sperm mitochondria after their entry in the unlabeled oocytes (Sato et al. 2018; Zhou et al. 2016; our unpublished data). The absence of TMRE in sperm mitochondria in TMRE-labeled embryos was also observed when SYTO11-labeled males were crossed with TMRE-labeled hermaphrodites. In this last experiment, only the oocyte-derived mitochondria were labeled with TMRE (Zhou et al. 2016). This result would argue for a loss of membrane potential of sperm mitochondria in the embryo rather than the dilution effect of the TMRE in an unlabeled embryonic environment. More experiments are needed to confirm this interpretation and to clarify the potential causal relationship between an early loss of membrane potential and their targeting by autophagy. It is also not clear yet what would be the mechanism responsible for the loss of the membrane potential. The loss of TMRE from the sperm-derived mitochondria was observed even in *allo-1* and *ikke-1* mutant embryos. This argues for the loss of potential being a very early step not depending on ALLO-1 recruitment nor IKKE-1 kinase activity (Sato et al. 2018).

2.2.2 Endonuclease mtDNA Degradation

Interestingly, in *C. elegans*, like many animal species, the 35–50 copies of sperm mtDNA (Al Rawi et al. 2011) are largely outnumbered in the embryo containing around 2.4×10^4 copies of maternal mtDNA (Tsang and Lemire 2002a). Despite the strong dilution of sperm mtDNA molecules, an active mechanism is required to degrade them in the embryo.

The mtDNA clearance in other animal species has been described before and after fertilization (Nishimura et al. 2006). In the fruit fly, it requires the activity of endoG, an endonuclease protein contributing to the degradation of mtDNA during spermatozoid individualization (Politi et al. 2014). This is coupled to a mechanical exclusion of mtDNA nucleoids during the individualization of the flagella. In endoG mutant flies, the mtDNA is stabilized.

In *C. elegans*, an RNAi screen against predicted mitochondrial proteins encoded by the nuclear genome coupled to a PCR-based assay to detect mtDNA stabilization revealed CSP-6, a mitochondrial endonuclease G, as a protein required for normal timing of sperm mtDNA clearance in the embryos (Fig. 7). These results were confirmed in a mutant expressing a CSP-6 with a deletion of its catalytic domain (Zhou et al. 2016). In embryos lacking a functional CSP-6 protein, a trackable form of sperm mitochondrial genome was found beyond the 64-cell stage up to the fourfold stage of embryo development (around 11 h of development). The observation of embryos from CSP-6-depleted hermaphrodite crossed with WT males with mitotracker-labeled mitochondria revealed that sperm mitochondria remain visible up to the coma stage (around 7 h of development). The apparent shift in stability between the sperm mtDNA and the labeled sperm mitochondria may be caused by several experimental differences with the PCR assay being a more sensitive technic to reveal residual sperm mitochondria material. The delay in sperm mtDNA and mitochondria degradation upon CPS-6 loss was associated with a slight delay in the autophagosomes enclosure around sperm mitochondria from the 4- to the 16-cell stages (Zhou et al. 2016).

2.2.3 Prohibitin 2 in Sperm Mitochondria Degradation

A sperm factor that could serve as a mark for sperm mitochondria targeting has been proposed (Fig. 7). The worm homologue of vertebrate prohibitin 2 (PHB2) mitochondrial protein is required for normal sperm mitochondria degradation (Wei et al. 2017). RNAi depletion of PHB 2 in adult males prior to their crosses delays sperm mitochondria degradation in the embryos. Sperm mtDNA was also stabilized in the F1 progeny from the cross involving *phb-2* RNAi-treated males (Wei et al. 2017). Interestingly, a proteasome-dependent rupture of the outer mitochondrial membrane is required for the mitophagy in HeLa cells and the interaction of LC3 with PHB2.

Surprisingly, while PARKIN and MUL1 were shown to be required for sperm-derived mitochondria mitophagy in mouse embryos (Rojansky et al. 2016), inactivation of the Pink/Parkin pathway in *C. elegans* did not prevent sperm mitochondria degradation. Indeed, paternal mitochondria labeled with HSP-6::GFP and MOs were still degraded in the *sqst-1* (p62 homologue), *pink-1*, and *pdr-1* (Parkin homologue) deletion mutants (Sato et al. 2018). This result indicates that the mechanism of degradation of sperm-inherited mitochondria is different from the classical mitophagy pathway targeting defective mitochondria. This is surprising since PHB2 is required for Parkin-mediated mitophagy in murine embryonic fibroblasts. More work is required to understand how *phb-2(RNAi)* depletion in sperm cells stabilizes the sperm mitochondria in the *C. elegans* embryo.

2.3 Impact of Oocyte-Derived Mitochondria Dynamics on Sperm Organelles Clearance

It has been suggested that mitochondria dynamics, this means a proper balance of fusion and fission of both maternal and paternal mitochondria, has an important effect in sperm organelles clearance after fertilization. Mitochondria continually change shape through the combined actions of fusion and fission (van der Bliek et al. 2013). In the case of mitochondrial fusion, tubular and elongated organelles are generated, while fission process generates fragmented ones. *C. elegans fzo-1* and *drp-1* genes encode GTPases of the dynamin family that are orthologues of the MFN1/FZO1 protein (Eura et al. 2003; Rolland et al. 2009), and the DRP1 protein (Labrousse et al. 1999), which are required for fusion and fission of mitochondria, respectively (Zamponi et al. 2018).

Loss-of-function mutations in these genes, *drp-1(tm1108)* which causes severe mitochondrial fission defects and *fzo-1(tm1133)* which affects normal fusion dynamics of mitochondria, allowed the study of mitochondrial network and morphology in *C. elegans* embryos and sperm (Wang et al. 2016). Using TMRE to specifically label mitochondria and electron microscopy analysis, Wang et al. showed that mitochondria in WT embryos display different elongated shapes and sizes and are broadly distributed in the cytoplasm. As anticipated, in mutants affecting mitochondrial dynamics, maternal mitochondria show radically different morphology and distribution. In *drp-1(tm1108)* embryos, asymmetric concentrated clusters of long and highly connected mitochondria are observed, while in *fzo-1(tm1133)* embryos mitochondria are mostly fragmented. Interestingly, in the *fzo-1(tm1133); dpr-1(tm1108)* double mutant, fusion defects caused by *fzo-1(tm1133)* is completely suppressed by the mitochondrial fission defect caused by *drp-1(tm1108)* leading to the formation of long and highly connected mitochondrial. Moreover, each mutation delayed the clearance of sperm mitochondria labeled with Mitotracker Red (MTR) in cross-fertilized embryos from WT males with *drp-1(tm1108)* as well as WT males with *fzo-1(tm1133)* hermaphrodites. This demonstrates that mitochondrial dynamics equilibrium in the embryo is important for paternal mitochondria elimination in *C. elegans*.

The impact of the mutations affecting mitochondrial dynamics was also tested on sperm mitochondria. While EM analysis of WT worms revealed mostly spherical mitochondria with a diameter of 0.5 μm, they appear larger in *fzo-1(tm1133) mutant* and larger but elongated in *drp-1(tm1108)* mutant spermatozoa. The authors observed that the paternal fission defect in *drp-1(tm1108)* males led to the stabilization of the sperm mitochondria up to the 500-cell stage, while sperm mitochondria from the *fzo-1 (tm1133)* males were destabilized with a total removal of these organelles at the 32-cell stage. In cross-fertilized embryos of *fzo-1(tm1133)* and *drp-1(tm1108)* worms, these opposite effects neutralize each other, and mitochondria are eliminated at a normal rate and are cleared in the 64-cell stage just as in the control.

The stabilization of sperm mitochondria in embryos coming from WT MTR-stained males and *fzo-1(tm1133)* hermaphrodites was consistent with the lack of autophagosome formation around sperm mitochondria in these zygotes. The quantification of the colocalization of immunolocalized LGG-1 and MTR-labeled paternal mitochondria revealed that fusion defects from the maternal side impair autophagosome formation around sperm mitochondria. Interestingly, once again this impairment is overcame in the *fzo-1; drp-1* double mutant hermaphrodite.

Additional TMRE and TEM experiments showed that *fzo-1(tm1133)* maternal mitochondria were targeted by the autophagy machinery, which led to think that a competition between maternal and paternal mitochondria for the autophagy machinery could be responsible for the delay of sperm organelles clearance in these zygotes.

3 Sperm Mitochondria Degradation to Prevent Heteroplasmy

3.1 Possible Consequences of Heteroplasmy

The strain containing the steady-state heteroplasmy with the *UaDf5*- truncated allele of the mtDNA was instrumental, once again, to evaluate the consequences of heteroplasmy brought by the spermatozoon upon fertilization (Tsang and Lemire 2002b). As mentioned in Sect. 1.1, this viable heteroplasmic strain shows reduced egg-laying and defecation rates, a shorter life-span, as well as a reduced motility of sperm cells (Liau et al. 2007). It also displays a higher embryonic lethality than homoplasmic wild-type worms (Zhou et al. 2016). In previous experiments, the stabilization of sperm mitochondria was achieved by bulk autophagy inactivation (Al Rawi et al. 2011; Sato and Sato 2011; Zhou et al. 2011) and lysosomal defects (Zhou et al. 2011), two cellular functions essential for worm development that are involved in many aspects of embryonic development (Palmisano and Melendez 2018). Furthermore, none of them is specifically interfering with the sole degradation of sperm mitochondria since they also delay the clearance of the MOs (Al Rawi et al. 2011; Sato and Sato 2011). It was therefore difficult to evaluate the impact on the progeny of the stabilized sperm mitochondria.

Interestingly, the delay in sperm mtDNA clearance has been, for the first time, associated with a detectable phenotype in the heteroplasmic embryos. A first cross experiment was done using heteroplasmic males containing sperm mitochondria with both the *UaDf5* and WT haplotypes (Zhou et al. 2016). Crossing these heteroplasmic males with CPS-6-deficient hermaphrodites produced a progeny with a higher level of embryonic lethality compared to the crosses using homoplasmic WT males (5.9 vs 0.7% embryonic lethality). The cell division timing of embryonic blastomers was doubled in crosses with the heteroplasmic males compared to the crosses using homoplasmic WT males. Significant but milder defects were also observed for the progeny of crosses between two *C. elegans* haplotypes from WT isolates when the males carrying the N2 haplotype were defective for CPS-6 activity (Zhou et al. 2016). This is the first indication that embryonic heteroplasmy induced by the stabilization of sperm-inherited mitochondria has deleterious effects with an increased lethality and cell cycle duration. It would be interesting to conduct the same phenotypical analysis in embryos with stabilized sperm mitochondria by other experimental means, for example, upon inactivation of the autophagy adaptor protein ALLO-1 (Sato et al. 2018). *allo-1* mutants are viable (Sato et al. 2018); therefore, it would be interesting to measure cell cycle duration as well as embryonic lethality.

3.2 *Probably More than One Mechanism to Prevent mtDNA Transmission*

Sperm mtDNA and mitochondria may be stabilized in the F1 progeny by several experimental means (see above). None of the methods used so far allowed obtaining a stable and inherited heteroplasmy. This is surprising since the trackable sperm mtDNA used was the deleted *UaDf5*, a molecule initially obtained after a mutagenesis, proving that it was able to colonize the worm. Nevertheless, since the total mtDNA content is stable until the third larval stage and the increase in mtDNA quantity is linked to the increase in the number of germ cells, it is likely that there is no mitochondrial DNA replication and this is therefore not a favorable cellular context to allow the amplification of the sperm-derived mtDNA. The sperm-derived mtDNA must reach the germ line lineage during embryonic development to get a chance to be amplified. In such situation, the smaller *UaDf5* mtDNA would have then a replicative advantage compared to the WT maternal mtDNA and might invade the germ line and be transmitted to the progeny. When the sperm mitochondria themselves are followed over time in the developing animal using mitotrackers or GFP fusions, they eventually disappear even in autophagy or lysosomal deficient animals. The same occurs with MOs. In these experiments, it is difficult to rule out that the structure do not simply loose the marker. The fusion of stabilized mitochondria with the maternal network might, for instance, dilute the marker enough to prevent their localization.

Furthermore, we are still facing experimental limits. For example, when the fluorescent markers like GFP::ANT-1 (Al Rawi et al. 2011) or GFP::HSP-6 (Sato and Sato 2011) were used to follow the sperm mitochondria, they were genetically encoded and expressed in the germ line by the paternal nuclear genome. This property rapidly precludes using them for the specific tracking of sperm mitochondria over generations since the marker is transmitted and expressed in the progeny. So far, these mitochondrial markers were expressed under germ line-specific promotors (*pie-1* or *mex-5* promotors) in order to load spermatozoa with the fluorescent-tagged protein. In these cases, the germ line expression starts again in the precursor germ cells in late embryogenesis, preventing to follow inherited sperm mitochondria in these cells after that stage. Regarding the MOs, it is very likely that they are still degraded in absence of functional autophagy machinery since none of the genetic background stabilizing the sperm organelles led to the transmission and accumulation of MO markers over time. For instance, ALLO-1 mutant is viable and would accumulate MOs over generation if they were stabilized and randomly segregated within the developing embryo including in the precursor germ cells. No such accumulation has been observed using an MO-specific antibody.

It is time to address the question of the mechanisms insuring sperm organelle clearance in a more integrated manner. Several mechanisms are important for normal sperm organelle degradation, but we still do not have a clear overview of how they integrate and what is the relative importance of these pathways in the normal *C. elegans* embryo. We now have more indications on how the embryo reacts to sperm organelles entry, but the signal(s) that allows their specific targeting remain elusive. *C. elegans* represents an appealing model to conduct unbiased genetic screens to identify mutants transmitting their sperm mtDNA and such possible sperm signal. Genetic screens should be conducted in (1) mutated maternal genetic backgrounds with a delay in sperm organelle degradation phenotype in order to identify the potential additional mechanisms besides autophagy and (2) on the paternal side in order to identify the sperm-specific signal for the degradation. The genetic approach could in principle reveal not only proteins but also posttranslational modifications or other types of signals like noncoding RNAs. Modern quantitative and comparative proteomic approaches of sperm and oocytes mitochondria composition could also reveal a specific factor involved in sperm mitochondria recognition.

Acknowledgments Our laboratory is supported by the Fondation pour la Recherche Medicale (Equipe FRM DEQ20160334874) and the European COST Program (BM1408 GENiE). We are grateful to Martin Sachse from Institut Pasteur Paris for the TEM and Charlène Perrois, Valeria Parrales, and Sebastien Normant for their technical support.

References

Al Rawi S, Louvet-Vallee S, Djeddi A, Sachse M, Culetto E, Hajjar C, Boyd L, Legouis R, Galy V (2011) Postfertilization autophagy of sperm organelles prevents paternal mitochondrial DNA transmission. Science 334:1144–1147

Alberti A, Michelet X, Djeddi A, Legouis R (2010) The autophagosomal protein LGG-2 acts synergistically with LGG-1 in dauer formation and longevity in C. elegans. Autophagy 6:622–633

Anderson S, Bankier AT, Barrell BG, de Bruijn MH, Coulson AR, Drouin J, Eperon IC, Nierlich DP, Roe BA, Sanger F, Schreier PH, Smith AJ, Staden R, Young IG (1981) Sequence and organization of the human mitochondrial genome. Nature 290:457–465

Chatterjee I, Richmond A, Putiri E, Shakes DC, Singson A (2005) The Caenorhabditis elegans spe-38 gene encodes a novel four-pass integral membrane protein required for sperm function at fertilization. Development 132:2795–2808

Djeddi A, Al Rawi S, Deuve JL, Perrois C, Liu YY, Russeau M, Sachse M, Galy V (2015) Sperm-inherited organelle clearance in C. elegans relies on LC3-dependent autophagosome targeting to the pericentrosomal area. Development 142:1705–1716

Eura Y, Ishihara N, Yokota S, Mihara K (2003) Two mitofusin proteins, mammalian homologues of FZO, with distinct functions are both required for mitochondrial fusion. J Biochem 134:333–344
Fawcett DW (1981) The cell. W. B. Saunders, Philadelphia
Hajjar C, Sampuda KM, Boyd L (2014) Dual roles for ubiquitination in the processing of sperm organelles after fertilization. BMC Dev Biol 14:6
Italiano JE Jr, Roberts TM, Stewart M, Fontana CA (1996) Reconstitution in vitro of the motile apparatus from the amoeboid sperm of Ascaris shows that filament assembly and bundling move membranes. Cell 84:105–114
Jacob WA, Bakker A, Hertsens RC, Biermans W (1994) Mitochondrial matrix granules: their behavior during changing metabolic situations and their relationship to contact sites between inner and outer mitochondrial membranes. Microsc Res Tech 27:307–318
Labrousse AM, Zappaterra MD, Rube DA, van der Bliek AM (1999) C. elegans dynamin-related protein DRP-1 controls severing of the mitochondrial outer membrane. Mol Cell 4(5):815–826
Lambert CC, Epel D (1979) Calcium-mediated mitochondrial movement in ascidian sperm during fertilization. Dev Biol 69:296–304
Lemmon MA (2008) Membrane recognition by phospholipid-binding domains. Nat Rev Mol Cell Biol 9:99–111
Liau WS, Gonzalez-Serricchio AS, Deshommes C, Chin K, Lamunyon CW (2007) A persistent mitochondrial deletion reduces fitness and sperm performance in heteroplasmic populations of C. elegans. BMC Genet 8:8
Luo SM, Ge ZJ, Wang ZW, Jiang ZZ, Wang ZB, Ouyang YC, Hou Y, Schatten H, Sun QY (2013) Unique insights into maternal mitochondrial inheritance in mice. Proc Natl Acad Sci U S A 110:13038–13043
Machaca K, L'Hernault SW (1997) The Caenorhabditis elegans spe-5 gene is required for morphogenesis of a sperm-specific organelle and is associated with an inherent cold-sensitive phenotype. Genetics 146:567–581
Manil-Segalen M, Lefebvre C, Jenzer C, Trichet M, Boulogne C, Satiat-Jeunemaitre B, Legouis R (2014) The C. elegans LC3 acts downstream of GABARAP to degrade autophagosomes by interacting with the HOPS subunit VPS39. Dev Cell 28:43–55
Melendez A, Talloczy Z, Seaman M, Eskelinen EL, Hall DH, Levine B (2003) Autophagy genes are essential for dauer development and life-span extension in C. elegans. Science 301:1387–1391
Nelson GA, Ward S (1980) Vesicle fusion, pseudopod extension and amoeboid motility are induced in nematode spermatids by the ionophore monensin. Cell 19:457–464
Nishimura H, L'Hernault SW (2017) Spermatogenesis. Curr Biol 27:R988–R994
Nishimura Y, Yoshinari T, Naruse K, Yamada T, Sumi K, Mitani H, Higashiyama T, Kuroiwa T (2006) Active digestion of sperm mitochondrial DNA in single living sperm revealed by optical tweezers. Proc Natl Acad Sci U S A 103:1382–1387
Okimoto R, Macfarlane JL, Clary DO, Wolstenholme DR (1992) The mitochondrial genomes of two nematodes, Caenorhabditis elegans and Ascaris suum. Genetics 130:471–498
Palmisano NJ, Melendez A (2018) Autophagy in C. elegans development. Dev Biol. https://doi.org/10.1016/j.ydbio.2018.04.009
Politi Y, Gal L, Kalifa Y, Ravid L, Elazar Z, Arama E (2014) Paternal mitochondrial destruction after fertilization is mediated by a common endocytic and autophagic pathway in Drosophila. Dev Cell 29:305–320
Roberts TM, Pavalko FM, Ward S (1986) Membrane and cytoplasmic proteins are transported in the same organelle complex during nematode spermatogenesis. J Cell Biol 102:1787–1796
Rojansky R, Cha MY, Chan DC (2016) Elimination of paternal mitochondria in mouse embryos occurs through autophagic degradation dependent on PARKIN and MUL1. elife 5:e17896. https://doi.org/10.7554/eLife.17896
Rolland SG, Lu Y, David CN, Conradt B (2009) The BCL-2-like protein CED-9 of C. elegans promotes FZO-1/Mfn1,2- and EAT-3/Opa1-dependent mitochondrial fusion. J Cell Biol 186:525–540

Sato M, Sato K (2011) Degradation of paternal mitochondria by fertilization-triggered autophagy in C. elegans embryos. Science 334:1141–1144

Sato M, Sato K, Tomura K, Kosako H, Sato K (2018) The autophagy receptor ALLO-1 and the IKKE-1 kinase control clearance of paternal mitochondria in Caenorhabditis elegans. Nat Cell Biol 20:81–91

Shakes DC, Ward S (1989) Mutations that disrupt the morphogenesis and localization of a sperm-specific organelle in Caenorhabditis elegans. Dev Biol 134:307–316

Sharpley MS, Marciniak C, Eckel-Mahan K, McManus M, Crimi M, Waymire K, Lin CS, Masubuchi S, Friend N, Koike M, Chalkia D, Macgregor G, Sassone-Corsi P, Wallace DC (2012) Heteroplasmy of mouse mtDNA is genetically unstable and results in altered behavior and cognition. Cell 151:333–343

Smith JR, Stanfield GM (2011) TRY-5 is a sperm-activating protease in Caenorhabditis elegans seminal fluid. PLoS Genet 7:e1002375

Sutovsky P, Moreno RD, Ramalho-Santos J, Dominko T, Simerly C, Schatten G (1999) Ubiquitin tag for sperm mitochondria. Nature 402:371–372

Tsang WY, Lemire BD (2002a) Mitochondrial genome content is regulated during nematode development. Biochem Biophys Res Commun 291:8–16

Tsang WY, Lemire BD (2002b) Stable heteroplasmy but differential inheritance of a large mitochondrial DNA deletion in nematodes. Biochem Cell Biol 80:645–654

Tsang WY, Lemire BD (2003) The role of mitochondria in the life of the nematode, Caenorhabditis elegans. Biochim Biophys Acta 1638:91–105

Tsukamoto S, Kuma A, Murakami M, Kishi C, Yamamoto A, Mizushima N (2008) Autophagy is essential for preimplantation development of mouse embryos. Science 321:117–120

van der Bliek AM, Shen Q, Kawajiri S (2013) Mechanisms of mitochondrial fission and fusion. Cold Spring Harb Perspect Biol 5:a011072. https://doi.org/10.1101/cshperspect.a011072

Wang Y, Zhang Y, Chen L, Liang Q, Yin XM, Miao L, Kang BH, Xue D (2016) Kinetics and specificity of paternal mitochondrial elimination in Caenorhabditis elegans. Nat Commun 7:12569

Ward S, Carrel JS (1979) Fertilization and sperm competition in the nematode Caenorhabditis elegans. Dev Biol 73:304–321

Ward S, Argon Y, Nelson GA (1981) Sperm morphogenesis in wild-type and fertilization-defective mutants of Caenorhabditis elegans. J Cell Biol 91:26–44

Washington NL, Ward S (2006) FER-1 regulates Ca^{2+}-mediated membrane fusion during C. elegans spermatogenesis. J Cell Sci 119:2552–2562

Wei Y, Chiang WC, Sumpter R Jr, Mishra P, Levine B (2017) Prohibitin 2 is an inner mitochondrial membrane mitophagy receptor. Cell 168:224–238.e10

Wolf N, Hirsh D, Mcintosh JR (1978) Spermatogenesis in males of the free-living nematode, Caenorhabditis elegans. J Ultrastruct Res 63:155–169

Xu XZ, Sternberg PW (2003) A C. elegans sperm TRP protein required for sperm-egg interactions during fertilization. Cell 114:285–297

Zamponi N, Zamponi E, Cannas SA, Billoni OV, Helguera PR, Chialvo DR (2018) Mitochondrial network complexity emerges from fission/fusion dynamics. Sci Rep 8:363

Zhang Y, Yan L, Zhou Z, Yang P, Tian E, Zhang K, Zhao Y, Li Z, Song B, Han J, Miao L, Zhang H (2009) SEPA-1 mediates the specific recognition and degradation of P granule components by autophagy in C. elegans. Cell 136:308–321

Zhou Q, Li H, Xue D (2011) Elimination of paternal mitochondria through the lysosomal degradation pathway in C. elegans. Cell Res 21:1662–1669

Zhou Q, Li H, Li H, Nakagawa A, Lin JL, Lee ES, Harry BL, Skeen-Gaar RR, Suehiro Y, William D, Mitani S, Yuan HS, Kang BH, Xue D (2016) Mitochondrial endonuclease G mediates breakdown of paternal mitochondria upon fertilization. Science 353:394–399

Doubly Uniparental Inheritance of mtDNA: An Unappreciated Defiance of a General Rule

Eleftherios Zouros and George C. Rodakis

Contents

Abstract We recount the basic observations about doubly uniparental inheritance (DUI) of mtDNA in bivalvian mollusks with an emphasis on those that were obtained from work in Mytilus and appeared after the review by Zouros (Evol Biol 40:1–31, 2013). Using this information, we present a new model about DUI that is a revised version of previously suggested models. The model can be summarized as follows. A Mytilus female either provides its eggs with the "masculinizing" factor S and the "sperm mitochondria binding" factor Z, or it does not. This property of the female is determined by two nuclear genes, S and Z, that are always in the on/on or the off/off phase. In fertilized eggs without factors S and Z the embryo develops into a female and the sperm mitochondria are randomly dispersed among cells following development. In fertilized eggs with factors S and Z, the first factor causes the cell to become eventually sperm and the second causes the sperm

E. Zouros (✉)
Department of Biology, University of Crete, Heraklion, Greece
e-mail: zouros@biol.uoc.gr; zouros@uoc.gr

G. C. Rodakis
Department of Biology, National and Kapodistrian University of Athens, Athens, Greece

P. Sutovsky (ed.), *Cellular and Molecular Basis of Mitochondrial Inheritance*, Advances in Anatomy, Embryology and Cell Biology 231,
https://doi.org/10.1007/102_2018_4, Published online: 14 January 2019

mitochondria to aggregate and anchor to the nuclear membrane by binding to a specific motif of the sperm-derived mtDNA. Factors S and Z are continuously co-synthesized and co-localized in the cell line from the egg to the sperm. The sperm mitochondria of the aggregate escape the mechanism that eliminates the cell's mitochondria before the formation of the sperm. The rescued mitochondria are subsequently packed into five mega-mitochondria in the sperm and are delivered in the egg.

Keywords Blue mussel, Doubly uniparental inheritance, Mitochondrial DNA inheritance, Mytilus

1 Introduction

The mitochondrial DNA (mtDNA) was introduced in population genetics, phylogeny, and evolution by Avise et al. (1979) and became soon the most powerful tool in the field (Avise 2004). By that time it was firmly established that animal mtDNA was maternally inherited (Huthcinson al. 1974; Hayshi et al. 1978), a phenomenon that became known as strict maternal inheritance (SMI). It was subsequently observed that the maternal inheritance was not strict. MtDNA heteroplasmy, i.e., the presence of two or more types of mtDNA in the same individual, was reported in many animals and in some of these the heteroplasmy was shown to be due to "leakage" of paternal mtDNA – the erratic escape from a mechanism that prevents the inheritance of paternal mtDNA (White et al. 2008). Fisher and Skibinski (1990) noted that this type of heteroplasmy was unusually high in the mussel *Mytilus edulis* and that male mussels were more often heteroplasmic than females. They suggested biparental inheritance as the most probable explanation. Hoeh et al. (1991) also observed a high level of heteroplasmy for highly diverged mtDNA molecules in Mytilus and suggested biparental inheritance. Zouros et al. (1992a, b) confirmed biparental inheritance by analyzing a set of mussel pair-matings. Soon it became obvious that the mussel heteroplasmy could not be a simple case of paternal mtDNA leakage. A clear explanation of male heteroplasmy and paternal mtDNA transmission in mussels appeared in two pairs of publications (Skibinski et al. 1994a; Zouros et al. 1994a, b; Skibinski et al. 1994b). These authors described a new system of mtDNA inheritance, which is known as "doubly uniparental inheritance" (DUI) (Zouros et al. 1994a).

The main features of DUI can be summarized as follows: The species *M. edulis* has two clearly distinguishable mtDNA molecules, namely F and M. The F is transmitted though the egg, and the M through the sperm. Somatic tissues of both sexes and female gonads are dominated by the F, and male gonads by the M. DUI was subsequently found in several species of bivalve mollusks, now numbering more than 100 (Gusman et al. 2016). The basic features described above hold for all these species, but minor differences may exist. Several reviews of DUI have appeared, each summarizing the status of our knowledge at the time the review was written (Zouros 2000; Breton et al. 2007; Passamonti and Ghiselli 2009; Zouros 2013). Here we give an account of the present status of our understanding of DUI, paying special attention to results that appeared after the aforementioned reviews. Most of these results refer to *Mytilus edulis – galloprovincialis,* a pair of species that have different yet overlapping geographical distributions, hybridize readily and have very similar F and M genomes. The reader who wants to get a full account of observational and experimental data about DUI should, at the minimum, consult the review of Zouros (2013) and the papers on Mytilus that appeared after this review – and are cited here. We use the occasion of this review to synthesize whatever we currently know into a comprehensive model. The strength of empirical evidence supporting the various elements of the model varies considerably and the likelihood that some parts of the model, or even the entire model, could be proved wrong remains high. But it appears to us that this is the best working hypothesis we may present at this moment.

2 Observations and Models

2.1 The F and M Content of Somatic and Germ Cell Lines of Mytilus

A key observation that came from mussel pair matings and also from broods of single females collected from the wild (Kenchington et al. 2002) was that there are three types of females with regard to the sex ratio of their progeny: females that produce almost exclusively daughters (daughter biased females), females that produce sons in high percentages (son-biased females), and females that could not be assigned statistically to one or the other of the two previous classes (females with unbiased sex ratio). The origin of sperm that fertilized the eggs did not affect this property of the female. Given that there are no secondary sexual characteristics in Mytilus, this classification of females was originally based on the presence or absence of the M genome in the gonads of their progeny (presence equated to maleness, absence to femaleness). It was subsequently observed from pedigrees of laboratory lines that the sex-ratio bias was inherited: daughter-biased mothers produced more often daughter-biased female progeny than male-biased mothers. These observations led to conclusion that in Mytilus sex is determined exclusively by the mother's nuclear genotype. From previous studies it was known that the Mytilus sperm contains five (and, rarely, four or six) mitochondria of large size (Longo and Dornfeld 1967). These "mega-mitochondria" are apparently the product of fusion of many "normal size" mitochondria (in a way analogous to that known to occur in the sperm of Drosophila; DeLuca and O'Farrell 2012) and can be easily seen under the microscope if stained with Mitotracker Green FM.

Cao et al. (2004a) examined the fate of sperm mitochondria in eggs from daughter-biased and son-biased mothers. They observed that in eggs that came from daughter-biased mothers the sperm mitochondria were randomly dispersed in the egg cytoplasm and in subsequent egg divisions they segregated randomly among the blastomeres. In eggs that came from son-biased mothers the five sperm mitochondria remained in close proximity to each other. In the two-cell stage all five mitochondria were found in the same blastomere and very close to the cleavage furrow that separated the two cells. In the four-cell stage the sperm mitochondria could be seen aligned in the inner part of the cleavage furrow of blastomere D, which is the blastomere that gives rise to the mesoderm and the germ line (Verdonk and van den Biggelaar 1983). This type of segregation and positioning close to the cleavage furrow continued in subsequent cell divisions and in several occasions, when the sperm mitochondria could still be seen, it was observed to persist as late as the trochophore stage and even in larvae of stage D, almost 72 h post-fertilization (Cao et al. 2004a). The two patterns of sperm mitochondria behavior in early embryogenesis were named "dispersed" and "aggregate." The aggregation is not very faithful. Occasionally, one or even two sperm mitochondria can be found, in apparently random position, in different blastomeres than the one that contains the aggregate.

There is nothing special about the distribution of either the F or the M mtDNA types in female or male somatic tissues (Fig. 1A, Ba). Occasional presence of M in female somatic tissues is expected from the fact that in eggs destined to develop into females the sperm mitochondria are not eliminated, but disperse randomly in the blastomeres. This expectation is consistent with the observation (see Zouros 2013 for references in Mytilus and the clam *Ruditapes philippinarum*). No regularity was found in the presence/absence or amount of the M genome among female tissues. Erratic presence of the M genome in male somatic tissues is also expected from the occasional break-away of mitochondria from the aggregate that is formed in eggs destined to become males. Presence of the M genome in male somatic tissues was observed more often than female somatic tissues, but it was not commoner in tissues of mesodermic origin, as would be expected if sperm mitochondria broke away from the aggregate after the formation of the mesoderm.

Of special interest is the question of whether the M genome that leaks into female embryos may find its way to eggs, as is shown – as a possibility – in Fig. 1. Ghiselli et al. (2011) failed to detect the M genome in highly purified collections of eggs from the venerid *Ruditapes philippinarum* and suggested that there must be a mechanism of elimination of M during oogenesis. There is no evidence of such a mechanism in Mytilus. Garrido-Ramos et al. (1998), Obata et al. (2007), and Sano et al. (2007, 2010) reported occasional presence of M in Mytilus eggs along with the dominant F. Consistent with this is the detection, in low frequency, of triplasmic individuals that carried one F and two M genomes. Of these, the most abundant was inherited from the sperm that the other was apparently present in minute amounts in the egg along with the F (Obata et al. 2007; Theologidis 2007).

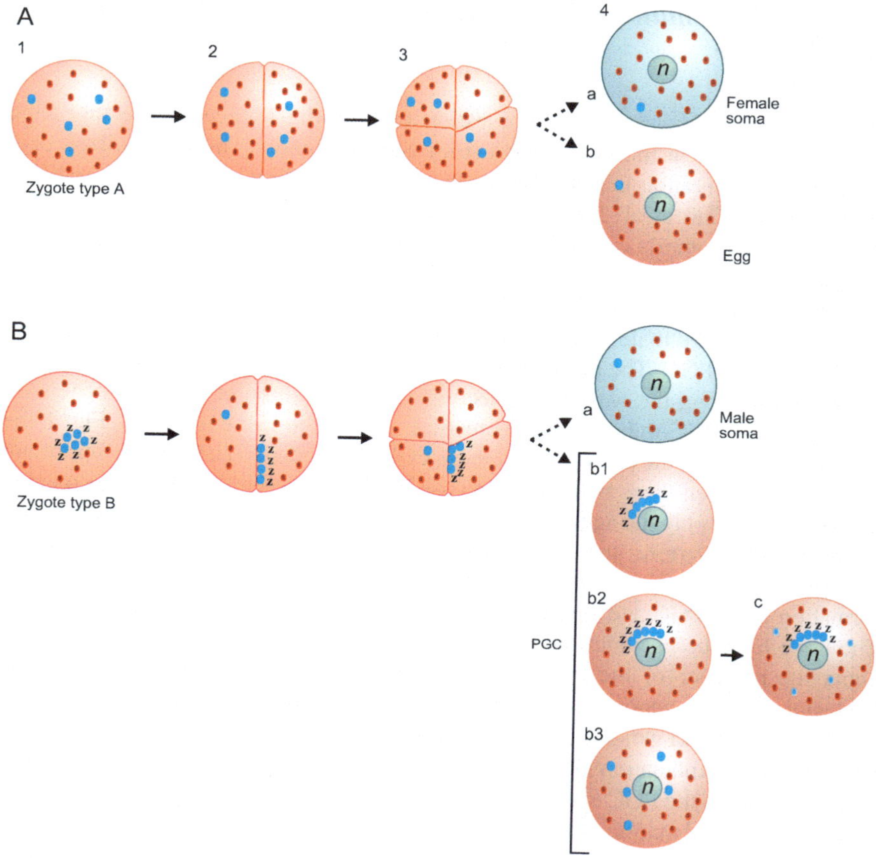

Fig. 1 Ontogenesis and gametogenesis in *M. edulis*. Blue dot: M-carrying mitochondrion, red dot: F carrying mitochondrion, n: nucleus, right-low cell: blastomere D. (A) Fertilized egg with the dispersed pattern of sperm mitochondria. The embryo develops into a female. A somatic cell may be free of M mtDNA or may contain it in tiny amounts. The latter possibility is even less likely for the egg. (B) Fertilized egg showing the aggregate pattern of sperm mitochondria. The embryo develops into a male. Z is the hypothesized factor responsible for the formation of the sperm mitochondria aggregate that binds to the mtDNA of sperm mitochondria and causes them to form an aggregate

The evidence that the sperm is free of the F genome is much stronger. Venetis et al. (2006) examined for the presence of F sperm of *M. galloprovincialis* that was not forced to swim through a solution of percol and sperm that was forced to swim. They detected the presence of F in 24 of 36 cases in the first and in none in the second. The study serves also as an illustration that contamination of DNA preparations from male gonads by the somatic cells containing the F is hard to avoid. Ghiselli et al. (2011) used also the percol assay and found no F in the sperm of *R. philippinarum*.

2.2 The Domination of the Male Germ Line by the M mtDNA Genome

We have no data bearing on the question of how the M genome becomes the dominant mtDNA in the male gonad from being a minority in the fertilized egg, beyond the observation that in Mytilus the sperm mitochondria aggregate remains until stage D (Cao et al. 2004a). Thus, we may only list alternative hypotheses (Fig. 1). One hypothesis is that the route of the sperm mitochondria that we described for the early egg divisions continues until the formation of the first primordial germ cell (PGC). We may call it "the PGC delivery" hypothesis to emphasize that there is a mechanism in the male embryo that directs and delivers the sperm mitochondria aggregate in the PGC. The mechanism does not prevent detachment or loss of sperm mitochondria along the way. But it secures that all or some of these mitochondria will end in the PGC rather than in somatic tissues or be lost entirely. In Fig. 1 the PGC delivery hypothesis is presented by cases b1 and b2. There is an important difference between these two cases. In case b1 it is assumed that there is a mechanism that allows only the aggregate to enter the PGC. In case b2 there is no exclusion of the F mtDNA from the PGC. Case b3 is the alternative of the PGC delivery hypothesis. According to it the sperm mitochondria aggregate is dissolved before the formation of the PGC and sperm mitochondria divide along with egg mitochondria. The three hypotheses about the formation of the PGC lead to different hypotheses about the presence of the F and M genomes in the male gonad. In case b1 the exclusion of the F genome from the PGC explains at once why the male gonad is dominated by the M genome and why the sperm contains only this genome. But case b1 requires an explanation about the mechanism of delivery of the sperm mitochondria aggregate in the PGC and the exclusion of the F genome from the PGC. Case b2 also needs an explanation of how the sperm mitochondria are delivered in the PGC, but it also requires an explanation of how the M becomes dominant in the gonad in the presence of the F genome and how the F genome is eliminated from the mature sperm. The last two questions apply also to case b3.

All versions that we have listed above have been mentioned as possibilities in the studies that dealt with the presence of the F genome in the male germ line (see Zouros 2013 for references). These studies produced unclear and, sometimes, conflicting results, which, in large degree, can be attributed to difficulty to eliminate contamination of mtDNA preparations from male gonads with somatic cells that are rich in F mtDNA. The quantitative study of Obata et al. (2011) of the amount and the degree of expression of the F and the M genomes in male gonads of *M. galloprovincialis* produced a picture of widely varying F/M ratios in various stages of male gonad development. Interestingly, these authors observed expression of the F type in spermatogonia and spermatocytes but not in spermatids and spermatozoa. At present the consensus is that the F genome is present in the male germ line and that its elimination from the sperm occurs at some late stage of sperm formation.

2.3 The Exclusion of the Maternal mtDNA from the Sperm

The elimination of mtDNA from the sperm is the rule in the animal kingdom. Assuming that the F genome corresponds to the "standard" type on animal mtDNA, its elimination from the sperm is the expected outcome. Sutovsky et al. (1999, 2000) have shown that the mechanism of mtDNA elimination from sperm in bovines and monkeys involves ubiquitination of the membrane of sperm mitochondria and their subsequent disposal. It is, therefore, reasonable to depart from the assumption that an analogous mechanism exists in Mytilus (and, by extension, to all species with DUI) and that the M genome is immune to this mechanism. As we have noted, there is good evidence in the literature for the presence of the F genome in the male germ line. If so, how its elimination from the mature sperm is achieved? One possibility is that the F or the M genome, or both, code for an RNA or a protein factor which is an indispensible part of the mechanism. The other possibility is that the mechanism involves a cytoplasmic factor which somehow recognizes and selectively destroys the F mtDNA carrying mitochondria or, inversely, recognizes and protects the M mtDNA carrying ones.

The existence of open reading frames (ORF) in both mtDNA types in Mytilus and in unionids favored the hypothesis of an mtDNA coded factor. The hypothesis got further support from the detection of a protein that corresponds to an ORF in the F genome of fresh water mussels (Breton et al. 2009, 2011a). The association of this protein with exclusion of the F genome from the sperm in unionids remains to be established. The possibility that F or M ORFs code for a protein or an RNA product in Mytilus is remote. Chatzoglou et al. (2013) studied the mRNAs coded by both the F and M genomes of *M. galloprovincialis*. They identified the mRNAs of all mtDNA coded proteins in the primary transcript and the points of its cleavage in eight monocistronic, one tricistronic, and one possible bicistronic product. But they failed to identify any mRNA sequence that could match an ORF in either the F or the M mtDNA. A more detailed study of the transcriptosome of the same species by Kyriakou et al. (2014a, b) also produced no evidence for a product coded by an ORF of the M or the F genome.

The shift of attention to the possibility that the mtDNAs of Mytilus may contain sequences that interact with a cytoplasmic factor (or factors) produced more encouraging results. Kyriakou et al. (2015) performed a series of electrophoretic mobility shift assays (EMSA) with extracts from male or female gonad as the protein-factor source and with DNA clones spanning the control region (CR) of the M and F genomes as binding factors. The control region (CR) of both genomes has been divided into three domains on the basis of primary sequence variability among F, among M or divergence between M and F genomes (Fig. 2). The first domain (in the 5′–3′ direction) is highly variable (thus called first variable domain, VD1), the second is highly conserved (conserved domain, CD), and the third is a short highly variable piece (VD2). Kyriakou et al. (2015) identified a sequence of 151 bp in the

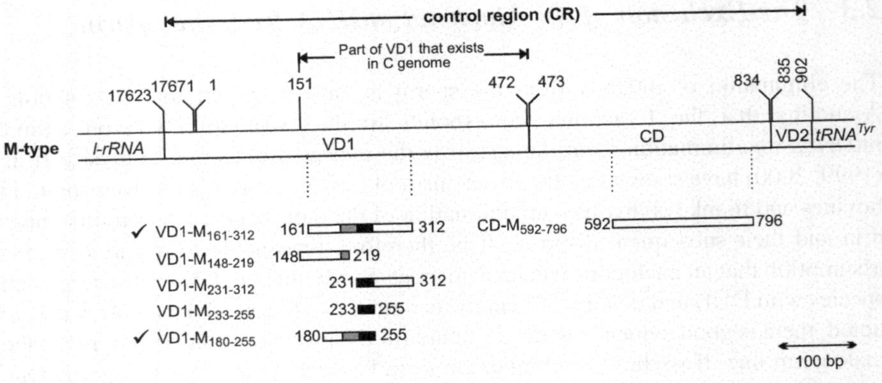

Fig. 2 Identification of a sequence motif in the M genome that interacts with a cytoplasmic factor. The main control region (CR) of the F and the M mtDNA of species of the genus Mytilus are flanked by the large ribosomal RNA (lrRNA) and the transfer RNA of tyrosine ($tRNA^{tyr}$). The lower part shows the clones that were used in EMSA assays. Dark box: the region that binds to protein factor; grey box: the polyA track that is necessary to bend the mtDNA for the facilitation of the binding; white box: a helper part of no-specific sequence. The check mark identifies the clones that produced the complex with proteins from the perinuclear cytoplasm. The smallest tested length of M mtDNA with the ability to produce the complex was 75 bp

VD1 region of the M genome that produced a shift with the extracts from the male gonad. Within this 151 bp segment there is a highly conserved element of 23 bp which is preceded by a poly-A sequence. Both the 23 bp element and the poly-A track are indispensable for the formation of the complex. The stretch upstream of the poly-A track is also needed for the formation of the complex, but need not have a specific sequence. Further, the authors experimented separately with nuclear and cytoplasmic extracts of the male gonad and found that the shift occurred with the nuclear but not the cytoplasmic extract. The shift was not produced when they used the same clones with extracts from eggs or from somatic tissues of males or females. Clones from the VD1 of the F genome produced no shift with extracts from male or female gonads. As a kind of control for the EMSA assay, Kyriakou et al. (2013) had previously used clones from the CD of the F and the M mtDNA. The CD contains sequences that elicit the molecule's replication and transmission. As expected, there was a shift with protein extracts from the cytoplasm.

The model proposed by Kyriakou et al. (2015) for the elimination of the F genome from the mature sperm is shown in Fig. 3. A cell that is destined to produce a spermatozoon is assumed to contain M-carrying and F-carrying mitochondria and a "protection factor," Z, which is located close to the nuclear membrane. Because the authors used for protein extraction whole male gonads (which consist of cells at various stages of maturation), cell 1 of Fig. 3 can, theoretically, be of any stage of

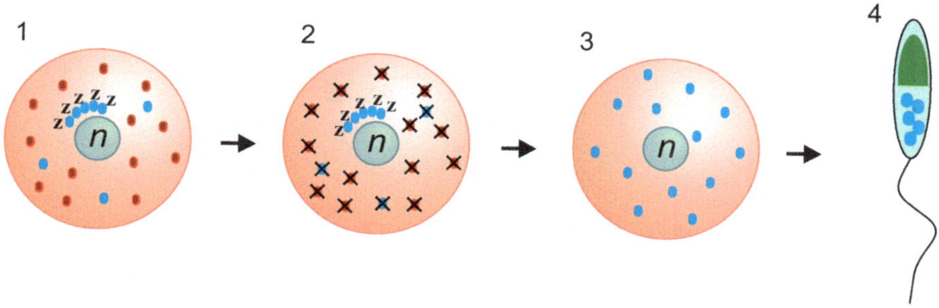

Fig. 3 A model for the exclusion of F mtDNA from the sperm. Blue dot: M-carrying mitochondrion, red dot: F carrying mitochondrion, n: nucleus, Z the hypothetical "protector" factor. Stage 1: Male gonadal cell. Stage 2: Gonadal cell just before it becomes spermatid. Stage 3: Spermatid. Stage 4: Mature sperm. Mitochondria marked with x denote mitochondria that are destroyed in late spermatogenesis by a mechanism that is assumed to be common to species with and without DUI. Z is the same factor shown in Fig. 1

maturation, from early descendants of PGC to just before spermatid formation. At stage 2, factor Z binds to M-carrying mitochondria that are around the nucleus. Again, the binding of factor Z with M-carrying mitochondria and the positional association of the complex with the nuclear membrane may have occurred at any stage in the development of the male gonad. At stage 2, a cytoplasmic factor (not shown) attacks and destroys the mitochondria irrespective of their DNA content, except those that have formed a complex with factor Z. F-carrying mitochondria are unprotected and destroyed and so are M-carrying mitochondria located away from the nucleus. The result is that spermatids (Fig. 3) and mature sperm (Figs. 3 and 4) contain only the M genome. This is in agreement with the results of Obata et al. (2011), who observed expression of the F type in spermatogonia and spermatocytes but not in spermatids and spermatozoa.

The model assumes that the mechanism of sperm mitochondria destruction (which is, apparently, present in all animals in one form or another) in Mytilus (and perhaps in all species with DUI, if not all bivalvian mollusks) acts at the pre-spermatid stage. There are two possibilities worth of consideration for the transition from the spermatid to the spermatozoon. The first is that the aggregate pattern of the five mega-mitochondria that is seen in the spermatozoon is formed concurrently with or immediately after the destruction of the non-protected mitochondria. The second is that rescued mitochondria divide and produce a large number of mitochondria from which the aggregate is formed along with the formation of the mature sperm.

parents

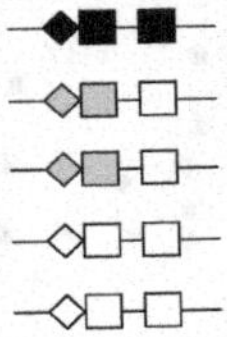

sex	genotype	gametes: sex	gametes: type	gametes: frequency
male	ee [R], Ee [H], EE [D]	sperm	(Z^-S^0)	[1]
female	ee [r]	eggs	(Z^-S^1)	[1]
female	eE [h]	eggs	(Z^-S^1)	[k]
			(Z^+S^2)	[1–k]
female	EE [d]	eggs	(Z^+S^2)	[1]

offspring

sperm	eggs: (Z^-S^1)	eggs: (Z^+S^2)
(Z^-S^0)	sex: female	sex: male
	mitochondria pattern: dispersed	mitochondrial pattern: aggregate
	soma: maternal mtDNA	soma: maternal mtDNA
	gametes: maternal mtDNA	gametes: paternal mtDNA

population

$h = [(1-(1-2k(1-k))^{1/2})/(2k(1-k))]$	$H = [(1-4d^2)/2]$
$d = [(1-2h(1-k))/2]$	$D = [(1+2d)^2/4]$
$r = [(1-2hk)/2]$	$R = [(1-2d)^2/4]$

Fig. 4 The genetic model for the explanation of the major features of DUI. The upper part gives the types of gametes produced by males and females of a different genotype at locus E. R, H, and D are the population frequencies of the three genotypes among males, and r, h, and d among females. Diamond is locus Z and squares are copies of locus S. Dark denotes unconditional inactivation, gray inactive state, and white active state. Superscripts of S denote the number of active copies of S, Z^- and Z^+ denote that locus Z is expressed or not expressed. The middle part gives the "phenotypes" of progeny resulting from any combination of sperm and egg. The lower part gives the equilibrium frequencies of the three genotypes at locus E as a function of *k*

2.4 *Maleness and Presence of Sperm-Transmitted mtDNA: An Associative But Not Causative Relationship*

As we have mentioned, sex determination in Mytilus is under the control of the female's nuclear genes and there is no role left for the male parent. Kiyomoto et al. (1996) induced triploidy in *M. galloprovincialis* and observed that all resulting individuals developed male gonads. Kenchington et al. (2009) treated eggs of *M. edulis* from mothers that produced almost exclusively daughters with cytochalasin B, which arrests meiosis at the first or second division, and examined whether (a) the sperm mitochondria aggregate was formed in fertilized eggs, (b) the resulting progeny developed a male or female gonad, and (c) the M genome was present in the gonad. They also examined no cytochalasin treated eggs from the same crosses for the same three characteristics. They observed that in treated eggs there was no aggregate formation, adult offspring from these eggs produced male gonads, and these gonads did not contain the M genome. Eggs from the same crosses that were not treated with cytochalasin B did not show the sperm aggregate and developed female gonads that lacked the M genome. They repeated the experiment with eggs derived from mothers that produced mainly sons. The majority of eggs showed the aggregate sperm in both treated and untreated eggs and developed male gonads that contained the M genome. The only difference between male gonads from untreated and treated eggs was that the gonad was normal in the first and with irregularities of different kinds in the second.

These results point directly to the following conclusions. Formation of the sperm aggregate and presence of the paternal mtDNA genome in the gonad of the resulting individual, and, reversely, lack of the aggregate and lack of the paternal genome, go together and are determined by the mother's nuclear genotype. Maleness is also determined by the mother's nuclear genotype, but it does not require formation of the sperm mitochondria aggregate or the presence of the paternal mtDNA genome in the gonad. Maleness is not dependent on the presence of the paternal mtDNA genome. Under normal circumstances (i.e., when the cross is homospecific and the eggs are not diplodized) maleness and sperm mtDNA co-occur. The simplest assumption of this coupling would be that the nuclear factor which is expressed in the female parent and causes maleness in her offspring and the factor that allows the sperm mtDNA to enter the germ line of these offspring are tightly linked in the nuclear genome and co-expressed in oogenesis. Triploidization interferes with sex determination but not with the fate of sperm mtDNA.

2.5 *"Masculinization" or Reversal of the Transmission Route of the F Genome*

The key observation that led to the discovery of DUI was the presence in any mussel population of two highly divergent mtDNA molecules of which one was transmitted though the egg and the other through the sperm. This picture was clouded by the detection of mtDNA molecules whose primary sequence highly resembled the sequence of the F genome, yet they were transmitted by the sperm. At the beginning these genomes were considered as F genomes which had "invaded" the male-transmitted mtDNA lineage and were, since then, transmitted by the sperm. They were also called "masculinized" genomes, in anticipation of "feminized" genomes, i.e., M genomes that might have invaded the female-transmitted lineage.

No feminized genomes were observed to this date. Masculinization presented, however, a major challenge for the understanding of DUI. Happily the problem was dissolved in a way that turned masculinization into a tool for the understanding of the molecular mechanism of DUI and its phylogenetic distribution in bivalve mollusks. One consistent observation in *Mytilus edulis/galloprovincialis* and the closely related species *M. trossulus* was that the masculinized genomes differed from the "true" (egg-transmitted) F genomes in the VD1 region of the control region (CR), thus making this region the primary suspect for sequences involved in the determination of the route of transmission (Burzynski et al. 2003; Cao et al. 2004b). Soon afterword it was shown that recombination between mtDNA molecules is possible in mussels (Ladoukakis and Zouros 2001) and that it had occurred in the history of several animal species (Piganeau et al. 2004; Tsaousis et al. 2005). Thus, the hypothesis that "masculinizing" sequences were transferred from the M genome to the F via recombination became very appealing. The conserved domain (CD) of the control region maintains the highest sequence similarity between the F and M genomes in Mytilus and could act as the site of the recombination.

The hypothesis was tested by obtaining the full sequence of the so-called C genome (Venetis et al. 2007). This is a masculinized genome that occurs in low frequencies in *M. galloprovincialis* populations, but it may reach high frequencies in some populations (Ladoukakis et al. 2002). Venetis et al. (2007) found that the protein and RNA coding parts of genome C differed little from the standard egg-transmitted F genome, but the CR was very different (Fig. 5). It consisted of four tandem repeats of the typical CR. The first and the fourth copy were mosaics of F and M sequences but the second and third were identical to the control region of the M genome with the exception of the first 199 bp of the VD1 domain that were missing. The most parsimonious explanation of the emergence of genome C is a duplication of the CR of the M genome which allowed for a misalignment of the M and F genomes and for a crossing over at the CD that shares high sequence similarities in the two genomes. Interestingly the compound CR of the C genome contains three identical sequences each of which contains the full 151 bp that Kyriakou et al. (2015) identified as the sequence necessary for the binding of the M genome with a factor in the perinuclear extract of male gonads (Fig. 2). The inference is that the C genome is rescued from the mitochondrial destruction that occurs in the pre-spermatid stage in the same way as the standard M genome.

2.6 *F/M Phylogeny and the Question of DUI Origin*

Figure 6 shows the phylogeny of 24 mtDNA genomes (13 F and 11 M) from fourteen DUI species with very different phylogenetic relationships. Several points of interest emerge. Within Mytiloida gender-specific clustering was observed only for the closely related species *M. edulis/M. trossulus* (which are clearly different species even though they may occasionally hybridize). The M genome of the cogeneric species *M. californianus* clustered away from its conspecific F and the four genomes of the other two species, reflecting the fast evolution of the M lineage. The two genomes of the other three mytilid species, *G. demissa*, *B. exustus*, and *M. senhusia*, clustered according to the origin of species. The same occurred with the venerid *V. philippinarum.* In contrast six F and four M genomes form six unionid

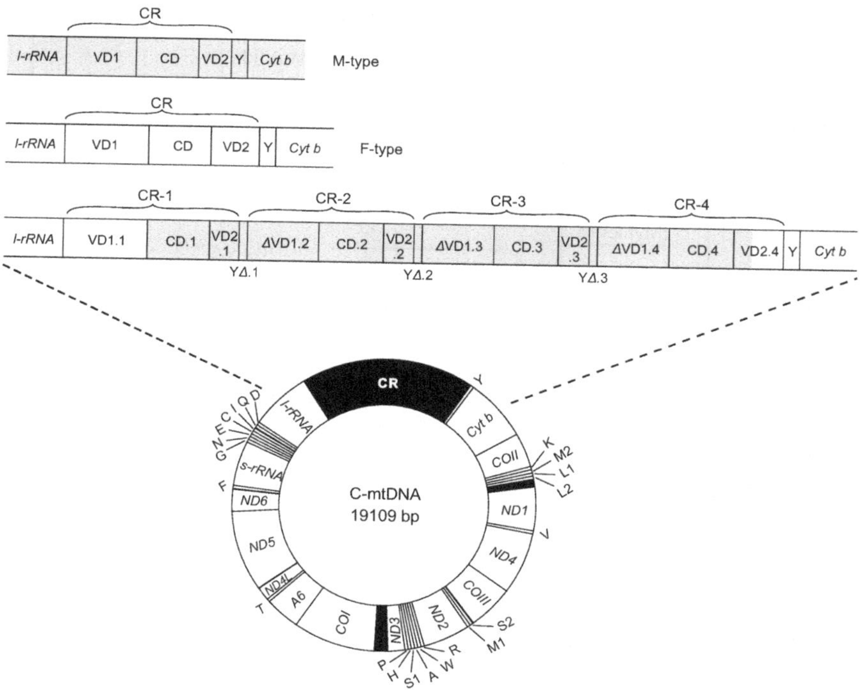

Fig. 5 The paternally inherited C mtDNA of *Mytilus galloprovincialis*. The entire genome is shown in circular form. Letters are the standard symbols for protein and RNA genes. Black boxes are replication and transcription control regions. The DNA sequence of all coding genes and of the two minor control regions are F-type. The CR of the C genome consists of three repeats of the CR of the M genome inserted between the VD1 and the Y gene of the F genome. Δ stands for deletion

species, some of which are as distantly related as the mytilid genera, formed an impressive gender-specific phylogeny.

A model that explains these observations is shown in Fig. 7. Part A shows three pairs of species of different phylogenetic relationship. It is assumed that the F and M genomes diverged from one common mtDNA in the distant past. Species 1 and 2 have recently diverged from a common ancestor species from which they inherited the same F and the same M genome. In time the F genomes that were "trapped" in different species diverged from each other and so did the M genomes, the latter at a higher rate. If scored at this stage the four genomes would cluster according to their mode of inheritance rather than taxon origin (gender-joining phylogeny). Given enough time a new sperm-transmitted genome, M3, will emerge from the F genome (first masculinization event) and may eventually replace M1. At the moment of masculinization the F/M divergence will be set to zero (the two genomes will differ only at the CR where the recombination that transferred the masculinizing sequences from the M to the F occurred). The descendant species 3 will contain this masculinized genome along with the F of species 1, which will have diverged and recognized

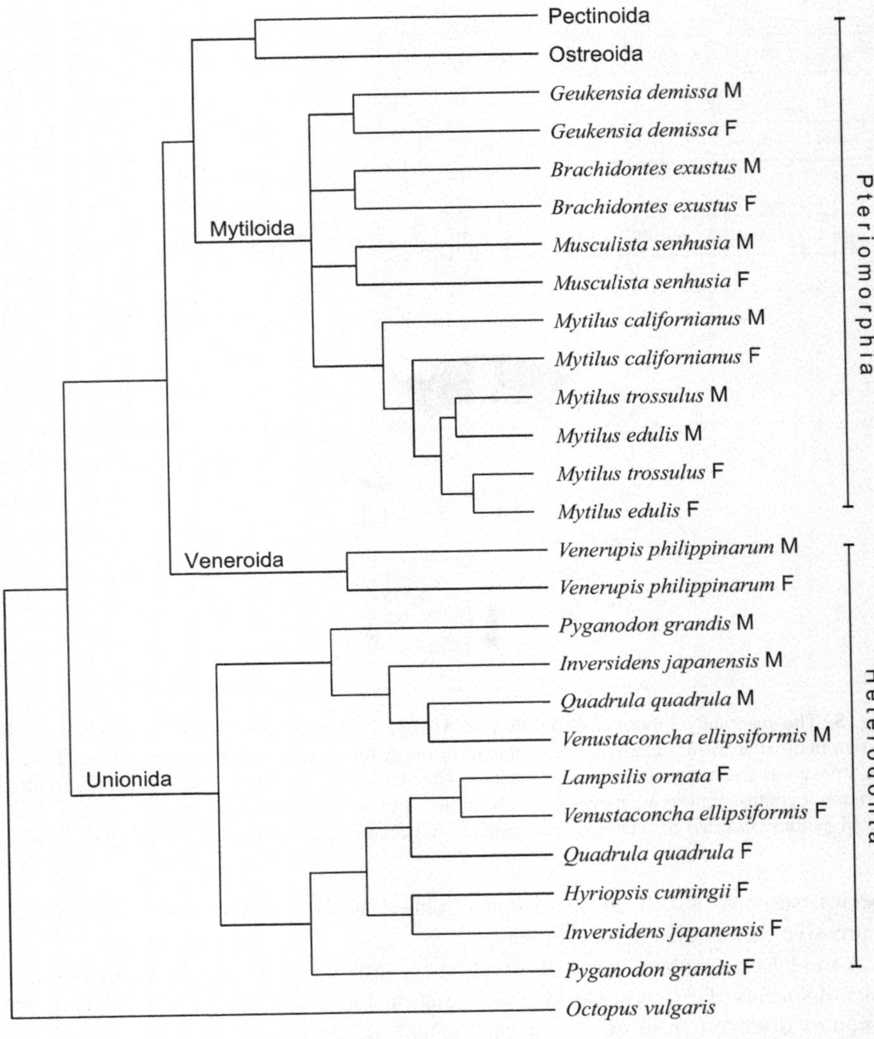

Fig. 6 The phylogenetic relationship of maternally (F) and paternally (M) transmitted mtDNA genomes of species from three super-families of bivalvian mollusks. Phylogenies involving a much larger number of species (Gusman et al. 2016) are consistent with the conclusions drawn from the above set of species (taken from Zouros 2013; original data from Theologidis et al. 2008 and Doucet-Beaupre et al. 2010)

as different (F3). If the four genomes of species 3 and 4 are grouped they will produce a tree that will reflect neither the mode of inheritance nor the taxon origin (mixed phylogeny). For two distantly related species (species 5 and 6) the time of divergence from the common ancestor species would be long enough for a masculinization to occur and replace the ancestral M. If this happened, the comparison of

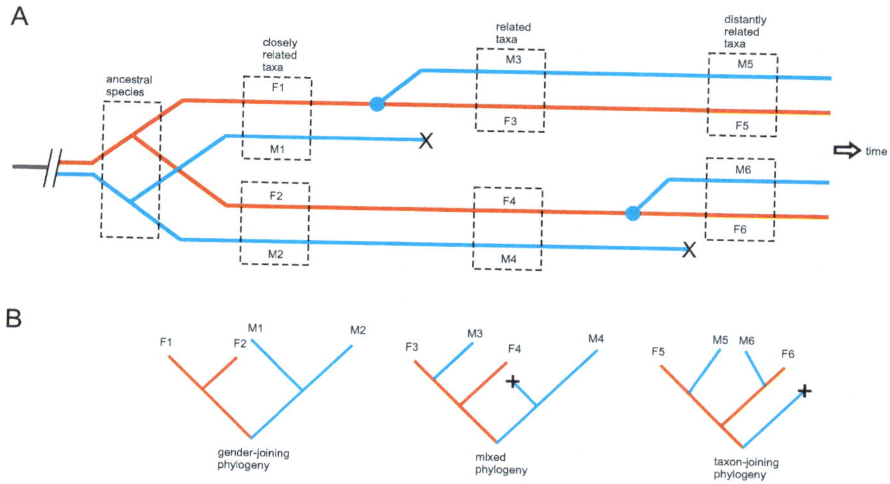

Fig. 7 How masculinization may explain the various phylogenetic patterns among F and M genomes from different taxa. (**a**) Three pairs of species of different phylogenetic distance. Dots denote masculinization events and X extinction events. (**b**) The three types of clustering of F and M genomes from a pair of species

species 5 and 6 would produce a taxon-joining phylogeny, i.e., one in which the two genomes of each species cluster together.

Figure 6 implies that multiple masculinization events followed by replacement of the old M have occurred in the long time (perhaps hundreds of millions years) that separated the deep branches of bivalvian mollusks, such as Pteriomorpha and Heterodonta. In Mytiloida we have all three types of phylogeny. There has been no fixed masculinization during the time that separated *M. edulis* from *M. trossulus*. One such event occurred after the lineage that produced *M. californianus* broke away from the lineage that produced the two previous species. All four mytiloid genera have been separated for a time long enough for multiple masculinization events to occur in the lineages that produced them. The simplest explanation for the gender-joining phylogeny of the unionid species is that in these species there can be no crossing over between the M and F genomes of the type that would transfer masculinizing sequences from the first to the second. This may be either because the conserved domains of the F and M phylads (the analogs of the CD of Mytilus) have diverged to the point that no recombination may occur or because the masculinizing sequences are not in the vicinity of conserved sequences.

The presence/absence of DUI in bivalve mollusks is spotty and irregular (Theologidis et al. 2008). On face value, this suggests either that DUI emerged once in the phylum of molluscan bivalves and was subsequently lost in several lineages or arose independently in several lineages. The evidence we have at present is not conclusive on this point. If DUI arose only once, the event must be very old, perhaps as old as 400 million years. If DUI has multiple and independent origins, its molecular mechanism must be fairly simple and only its details may vary among distant species that share the phenomenon. Such differences have, indeed, been observed. One is the aforementioned lack of masculinization in unionids and the amount of the M genome in somatic tissues. But, again, such variations may be expected under the hypothesis of single origin and the very long time that separates the major phylogenetic divisions of the super-class Autolamellibranchia that contains all known species with DUI. An answer could be obtained if the presence of DUI was found to co-occur consistently with some other basic feature of the species' biology. Breton et al. (2011b) observed discordance between DUI and hermaphroditism in a small collection of unionid species in which hermaphroditism appeared to have arisen independently. They suggested that the transition from dioeciousness to hermaphroditism is accompanied with DUI loss. The idea that hermaphroditism is incompatible with DUI is interesting and must be looked in more species with different taxonomic relationships.

The one-factor model that we describe here (see Sect. 2.8) has a strong bearing on the question of the origin of DUI. In a species that has the sex-determining mechanism that we proposed for Mytilus all that is required for the appearance of DUI is the emergence of a gene, Z, in close proximity to the on/off copy of the sex determining locus S (Fig. 4). This would mean that a species in which females are either daughter or son producers is vulnerable to becoming a DUI species. The inverse does not follow. Whereas any species with DUI will have females with highly biased sex-ratio progeny, not all species with this feature will have DUI. Under the one-factor model the multiple origin of DUI may not be highly improbable and two or more species with DUI need not be similar in all details. Loss of DUI is more problematic. If high sex-ratio biases of opposite direction among females is a condition for the operation of DUI, then loss of the bias in a species, for whatever reason, would mean the collapse of DUI and the transition to standard uniparental transmission of maternal mtDNA. Also, high sex ratio bias would elevate founder and small population size effects, making, in the long run, DUI species more vulnerable to extinction compared to species with standard maternal mtDNA inheritance. This might be one way to explain the presence of a species with standard maternal inheritance in a cluster of species or families that contain mostly DUI species.

2.7 *Why Is the M Genome Necessary for Male Fertility?*

As we have argued in a previous section, a specific short track of noncoding sequences from the M genome is capable to turn an F genome from female-transmitted to male-transmitted. Far from vanishing, these newly masculinized F genomes may replace eventually the old M from the entire species. This suggests that there is nothing the M genome does that the F genome cannot do, except to ride with the sperm. The short DNA sequence that the F genome "extracts" from the M is enough for the newly masculinized F to make the donor M superfluous for the species. But just riding by the sperm cannot be the only thing the M genome does. If that was the case, then an "old" M and a "new" M ought to be equal in their performance as sperm riders, and there would be no explanation of the replacement of the old M by the new in the population. In fact the evidence is that, barring extinction by random drift, the fixation of the new M is the expected outcome.

We have to conclude that the binding of the M genome to factor Z is not the only function of the M genome. This is clear from the fact that all the information we have about the protein and RNA coding parts of paternally transmitted genomes is that these parts are in functional order. The M genome is not loaded with pseudogenes as would be expected if only the masculinizing sequences of the VD1 were needed for its persistence in the population. What could this other function be? The answer becomes clear from inspection of Fig. 3. After the destruction of mitochondria in the pre-spermatid stage, there would be no mitochondria to fuel the process of sperm maturation other than the rescued M-carrying mitochondria. The M genome becomes indispensible for male fertility because there is no F genome to carry the process of spermatogenesis to the end. This explanation is consistent with the observations of Dalziel and Stewart (2002) and Obata et al. (2011). The latter observed no expression of the M genome in somatic tissues. In the male gonad they observed strong expression of the F genome in the pre-spermatid stages but not in spermatids and spermatogonia, in which there was strong expression of the M genome. These observations suggest that the rescued sperm mitochondria become detached from factor Z and become functional (case 3 in Fig. 3). This might not be the only stage in the life of the male mussel that the M genome is functional. For example, functionality of the M genome would be required if alternative b1 in Fig. 1 is the true case. But, as we have noted, this is highly unlikely.

There appears to be little doubt that the M genome plays a role in the development of the male gonad, but masculinization tells us that whatever that role is, it can be also fulfilled by the F genome. In fact, the repeated replacement in natural populations of M genomes by new F-like masculinized genomes suggests that this role can be played more efficiently by the F genome. Between two males that carry the same maternal genome but one carries an "old" paternal genome and the other a newly arisen one, the second must leave more progeny. This superiority should be attributed to the accumulation of mutations in paternally transmitted genomes that are not fully deleterious but reduce the efficiency of the genome in comparison to maternally transmitted genomes. Again, this is consistent with the faster evolution of the M genome, which is a common observation in DUI species. Indeed, paternally inherited genomes may be among the fastest evolving mtDNA molecules we know (Hoeh et al. 1996). The fast rate of divergence affects all protein-coding genes of the M genome, even though at a different rate, and suggests that it affects the basic function of the mitochondrion as an energy fueling apparatus of the cell. This explanation was proposed by Stewart et al. (1996) who interpreted the different rates of divergence of the two genomes in terms of selective constraint. The maternal genome has to fuel the somatic tissues of females and males, the female germ line and perhaps the male germ line before its elimination at the latest stages of sperm formation. In contrast, the paternal genome is needed only for the development of the male gonad, perhaps together with the F in early stages and alone after the formation of spermatids. This multiplicity of roles for the F genome, and particularly that of fueling the somatic tissues of both sexes, places a heavier constraint on its accumulation of mutations.

2.8 The One-factor Model for DUI

Zouros (2013) presented a model that captured the basic observations known when that review was written. According to the model, sex (which in Mytilus is equated to development of an egg or sperm producing gonad) is determined by a locus, S, that occurs in two copies, the proximal and the distal. One active copy produces females and two or more copies produce males. Both copies are silenced in males. The distal copy is always active in females, but the proximal copy can be active or not. The activation of the proximal copy depends on locus E that segregates for two alleles, E and e. In females of genotype EE the proximal copy of S is active, in females of genotype ee it is inactive, and in females of genotype Ee it is inactive at a rate *k* and active at a rate 1 – *k*. This system of sex determination is assumed to have preceded DUI.

The presence of paternal mtDNA in the gonad is controlled by locus Z. If locus Z is active the paternal mtDNA is present in the gonad, if it is inactive the paternal mtDNA is absent. Locus Z is tightly linked to the proximal copy of S and under the same expression regime, meaning that the proximal copy of S and Z are always in an on/on or off/off combination. According to the model the progeny of mothers of genotype ee will have one active copy of S and will develop into females. Their eggs will show the dispersed pattern of sperm mitochondria and the gonad will not contain the paternal mtDNA. The exact opposite is expected for mothers of genotype EE. In mothers of genotype Ee fraction *k* of eggs will show the dispersed pattern of sperm mitochondria and will develop in females with no paternal mtDNA in their gonads, and fraction 1 – *k* will show the aggregate pattern and will develop into males with the paternal mtDNA in their gonads. The model accounts for the unconditional maleness of triploids. Diploid eggs from mothers of any genotype at locus E would have at least two active S copies, so they would develop into males. Diploid eggs from zz mothers will have no active Z factor and when fertilized they will show the dispersed pattern, from ZZ mothers will have two active copies of factor Z and will show the aggregate pattern, and eggs from Zz mothers will either show or not show the aggregate pattern depending on the number of active Z copies they carry. Maleness and inheritance of paternal mtDNA are decoupled.

The above model (Zouros 2013) departed from assuming the existence of a factor, named factor X, that was present in all eggs and its role was to destroy the mitochondria of the sperm. As a second step, it introduced factor Z that was present only in eggs destined to become males. The role of Z was to suppress X and thus rescue the sperm mitochondria from destruction. This scheme was proposed before the work of Kyriakou et al. (2015) who suggested that a factor, named factor W, binds to the M genome and recues mitochondria that carry this genome from a mechanism that destroys the mitochondria before sperm formation (Fig. 3). This finding suggests that the mechanism that prevents the entrance of sperm mitochondria in the female embryo does not act after but before sperm formation. In turn this means that factor X does not exist and that the role of factor Z cannot be to rescue the sperm mitochondria from destruction in the egg but rather to protect the aggregate of sperm mitochondria from being dissolved.

The removal of the mitochondria destruction mechanism from the fertilized egg to the pre-spermatid stage is consistent with the observation that sperm mitochondria that failed to remain attached to the aggregate are not destroyed but can be seen to persist in the "wrong" blastomere. It also explains why the M genome is occasionally found in females. Most importantly, it reduces the three-factor (X, Z, W) DUI model into two-factor (Z, W) model: factor Z which allows sperm mitochondria to remain in aggregation and factor W which protects M-carrying mitochondria from the destruction mechanism that acts before the formation of the sperm. The two factors occur always in either the on/on or the off/off phase. This raises the possibility that the two factors are under the same expression regime or are in fact just one and the same factor, which we may call factor Z (the "one factor" model of DUI). The implication of the "one factor" model is that the mechanism that causes the sperm mitochondria to remain together in the fertilized egg does not differ from the mechanism that protects M-carrying mitochondria from being destroyed in the pre-spermatid stage.

Is it possible to explain the next steps in the fate of the aggregate without introducing a new factor? We may assume that the co-expression of Z with the proximal copy of S, which is responsible for the masculinization of the gonad, continues in the germ line but not in the soma (which would be consistent with the fact that there are no secondary sexual characteristics in Mytilus). This would mean that the binding of factor Z with the M genome is in operation in the male gonad from the moment of egg fertilization to the stage the F mtDNA carrying mitochondria are excluded from the pre-spermatid stage (Fig. 3). Factor Z is continually synthesized and, as M-carrying mitochondria divide, it binds to those that are in the vicinity of the nucleus and attaches them to the nuclear membrane. This scheme will explain the delivery of sperm mitochondria in the PGC and their persistence in the gonad. But requires the explanation of two other observations: (a) how does the aggregate recognize the cell during embryo development that would produce the PGC? and (b) how does it explain the explosion of the paternal mtDNA in the gonad? The answer to both questions must be sought at the co-expression of factor Z with the proximal copy of S that is responsible for maleness. The mechanism of formation of somatic and germ cell lines must impose a correlation between the positioning of the products of locus S (of both, the distal and the proximal copies) and the blastomere that would ultimately give rise to PGC (male or female). Co-transcription of factor Z with the proximal copy of S could have as a result that the functional products of the two loci stay in close proximity to each other and near the nuclear membrane for the entire time from the moment of fertilization that produced the male individual till the time the male's spermatids are formed. If development of a male gonad is dependent on high levels of expression of the proximal copy of S this would also mean high expression of Z. This in turn would mean that Z-bound mitochondria would become more numerous, leading to a disproportional increase of the amount of the M genome compared to that of F.

We have arrived at a "one-factor" model for DUI that allows for the presence of the maternal genome in the male gonad, in agreement with various studies, and one that explains most of what we know about the paternal genome, notably the behavior of sperm mitochondria from fertilization to early embryonic divisions in male embryos and the reaction of a perinuclear factor with M-containing mitochondria. Schematically the model is shown in Figs. 1 and 3. In Fig. 1, part A remains valid. In part B, case b2 is the preferred case for PGC formation. The work of Obata et al. (2011) suggests that the male gonad consists of cells that are similar to b2. Until formation of the spermatid these cells contain in constant flux sperm mitochondria that are bound to factor Z and are, apparently, not functional. Alternatively, case c depicts a cell of the male gonad that contains Z-bound sperm mitochondria and free sperm mitochondria that may be functional. At present we do not have enough information to choose between these two alternatives.

3 Some Outstanding Questions

The one-factor model can be easily disproved. For this it would be enough to show that the EMSA reaction detected by Kyriakou et al. (2015) does not occur in fertilized eggs produced from son-biased mothers. Inversely, if the reaction occurs in fertilized eggs produced from son-biased mothers but not in eggs from daughter-biased mothers, then the assertion that factors Z and W are one and the same thing would win credence and would bring us closer to the understanding of DUI. To gather further support, the test can be done with gonads arrested at an early stage of maturation and with somatic tissues. Clearly, this type of tests deserves top priority.

Immediately following in importance is to gain information about factor Z by making use of the tools of molecular biology and protein research. Are there homologous DNA sequences in non-DUI species, do they produce a product, and is the product involved in sex determination or mtDNA transmission? Clearly, coming to know more about factor Z would give answers to many of the peculiarities of DUI. At the same time it would set light on the question of one versus multiple DUI origins. It would make sense to assume that a precursor nuclear sequence existed prior to the evolution of the binding motif of the paternally inherited mtDNA. The mutability of the VD region is high enough for a motif with a binding ability to emerge by chance alone. So the hypothesis “Z first, binding motif second” appears much more likely than the inverse.

Going back in time, the question of a single origin versus multiple origins is of interest if for no other reason than that single origin would make DUI a very old phenomenon, even older than 400 million years. Multiple acquisitions and loses of DUI would mean major reorganizations of the species’ mode of reproduction with a high likelihood of extinction. If we adopt the reasoning of selfish elements, DUI can be seen as an excellent case of revolt of the paternal mtDNA against the despotism of the maternal mtDNA. But the selfishness of the paternal mtDNA comes at a cost. It requires the ability to do for the male gonad (at least at some stage in the maturation of the gonad) all that the maternal mtDNA does in species without DUI. This means that it should carry a complete set of coding genes, which in turn makes it vulnerable to mutation accumulation and eventual replacement by a molecule with a much lighter mutational load. Selfishness carries the punishment of much shorter phylogenetic life than the maternally inherited genome.

Another question that can be settled in Mytilus is whether there can be sperm heteroplasmy. This would be expected in the sperm of rare triplasmic males, i.e., males that resulted from eggs that contained an F and an M genome and would, therefore, have two M genomes, one maternal and one paternal. The test would require scoring percol-treated sperm from a large number of males at a highly variable site of the M genome, such as VD1. The results of Kyriakou et al. (2010) argue against sperm heteroplasmy. But, given that females with two M genomes are rare, the work must be repeated with sperm from a much larger number of males to elevate the probability that one or more of these males would be sons of a mother with two M genomes.

DUI has been observed in more than 100 species of bivalve mollusks. From most of them all that is known is the presence of two highly divergent mtDNA molecules of which one is present in males and absent, or rarely found, in females. The phenomenon has been studied in more detail only in three families, Mytilidae, Unionidae, and Veneridae. One reason for this is that species with DUI are not easy material for laboratory work. Yet it is important that the basic features that were observed in the aforementioned families be verified in other species. Most important among these is whether strong sex biases of opposite signs occur among females of species with DUI. This is confirmed in mytilids and the venerid *Ruditapes philippinarum* (Ghiselli et al. 2012). Machordom et al. (2014) examined broods from several females of the unionid *Unio delphinus* (Unionidae). Their results also pointed to, but did not provide good statistical support for, the presence of sex bias in this species. Ideally, we need to know if the bias exists in all species that are currently known to have DUI. This may not be a difficult task provided that one could collect and spawn a reasonable number of females, along a few males, from the species in question. The recognition of two different patterns in the distribution of sperm mitochondria in fertilized eggs would be the fastest and strongest test. Scoring for the presence of the paternal mtDNA among juveniles might be a second choice.

The extensive study of sex-ratio bias in Mytilus by Kenchington et al. (2002) suggested that daughter-bias was more extreme than the son-bias (which was the reason these authors used the term "sonless" females). This suggests that activation of the proximal copy of locus S, which needed for maleness, may not be complete in females of genotype EE (the male-biased mothers), and that it would vary widely in females of genotype Ee (Fig. 4). Variation in the value of k, either among broods of the same female and/or among different females, could be a way of response to needs for adjustment of the male/female ratio in the population. Simple calculations with different k values using the equilibrium genotype frequencies (bottom part of Fig. 4) show that small changes in k cause strong changes in the frequencies of the two homozygotes, ZZ and zz, in both sexes. If the above holds, it would be easy to study sex ratio responses in Mytilus by playing with various environment conditions, such as water temperature, salinity, or food availability.

4 Concluding Remarks

In species with DUI two different mtDNA molecules coexist stably and are transmitted from generation to generation with the same regularity and fidelity the X and the Y chromosomes coexist and are transmitted in a mammalian species. As far as we know, no DNA-carrying cytoplasmic organelle shows such a stable dimorphism with such a faithful mode of inheritance. Yet, we barely know anything about the molecular basis of the phenomenon. To use the X/Y analogy, we are at the stage when the role of this pair of chromosomes in sex determination was just discovered. Practically all the information we have about DUI is the outcome of the work of a handful of labs. DUI may well be a crying exception to the most general rule in the transmission of DNA-carrying cytoplasmic organelles, and more specifically of the animal mtDNA. Exceptions are, without doubt, signals to the underpinnings of the rule – the importance of Drosophila mutants in the foundation of Genetics is a typical example. There are good reasons why DUI failed to excite the interest of researchers in the field. The phenomenon appears to be restricted to a group of animals remotely related to humans. Even pure curiosity cannot be served well by these organisms because of the difficulty they present as experimental material. What are the prospects that DUI would open new insights in the role of mtDNA on modes of reproduction, sex inheritance, fertility, or allocation of reproductive effort between sexes? How would it contribute to the increasing list of hitherto unexpected functions of mtDNA or serve the quest for a deeper understanding of mtDNA-nuclear interactions and co-evolution? This is only a sample of questions one may ask, but apparently the time for the search of answers has not come.

Acknowledgment We thank Dr. M. Ladoukakis for offering helpful comments on an earlier version of the paper.

References

Avise JC (2004) Molecular markers, natural history and evolution. Sinauer Associates Inc., Sunderland, MA

Avise JC, Lansman RA, Shade RO (1979) The use of restriction endonucleases to measure mitochondrial DNA sequence relatedness in natural populations. I. Population structure and evolution in the genus Peromyscus. Genetics 92:279–295

Breton S, Doucet-Beaupre H, Stewart DT, Hoeh WR (2007) The unusual system of doubly uniparental inheritance of mtDNA: isn't one enough? Trends Genet 23:465–474

Breton S, Doucet-Beaupre H, Stewart DT, Piontkivska H, Karmakar M, Bogan AE, Blier PU, Hoeh WR (2009) Comparative mitochondrial genomics of fresh water mussels (Bivalvia:Unionoida) with doubly uniparental inheritance of mtDNA: gender-specific open reading frames and putative origins of replication. Genetics 183:1575–1589

Breton S, Ghiselli F, Passamonti M, Stewart DT, Hoeh WR (2011a) Evidence for a fourteenth mtDNA-encoded protein in female-transmitted mtDNA of marine mussels (Bivalvia: Mytilidae). PLoS One 6:e19365

Breton S, Stewart DT, Shepardson S, Trdan RJ, Bogan AE, Chapman EG, Ruminas AJ, Piontkivska H, Hoeh WR (2011b) Novel protein genes in animal mtDNA: a new sex determination system in freshwater mussels (Bivalvia: Unionoida)? Mol Biol Evol 28:1645–1659

Burzynski A, Zbawicka M, Skibinski DOF, Wenne R (2003) Evidence of recombination of mtDNA in marine mussel *Mytilus trossulus* from the Baltic. Mol Biol Evol 20:388–392

Cao L, Kenchington EL, Zouros E (2004a) Differential segregation patterns of sperm mitochondria in embryos of the Blue Mussel (*Mytilus edulis*). Genetics 166:883–894

Cao L, Kenchington EL, Zouros E, Rodakis GC (2004b) Evidence that the large noncoding sequence is the main control region of maternally and paternally transmitted mitochondrial genomes of the marine mussel (Mytilus spp.). Genetics 167:835–850

Chatzoglou E, Kyriakou E, Zouros E, Rodakis GC (2013) The mRNAs of maternally and paternally inherited mtDNAs of the mussel *Mytilus galloprovincialis*: start/end points and polycistronic transcripts. Gene 520:156–165

Dalziel AC, Stewart DT (2002) Tissue-specific expression of male-transmitted mitochondrial DNA and its implications for rates of molecular evolution in *Mytilus* mussels (Bivalvia: Mytilidae). Genome 45:348–355

DeLuca SZ, O'Farrell PH (2012) Barriers to male transmission of mitochondrial DNA in sperm development. Dev Cell 22:660–668

Doucet-Beaupre H, Breton S, Chapman EG, Blier PU, Stewart DT, Hoeh WR (2010) Mitochondrial phylogenetics of the Bivalvia (Mollusca): searching for the origin and mitogenomic correlates of doubly uniparental inheritance of mtDNA. BMC Evol Biol 10:50

Fisher C, Skibinski DOF (1990) Sex-biased mitochondrial heteroplasmy in the marine mussel Mytilus. Proc R Soc Lond B Biol Sci 242:149–156

Garrido-Ramos MA, Stewart DT, Sutherland BW, Zouros E (1998) The distribution of male-transmitted and female-transmitted mitochondrial DNA types in somatic tissues of blue mussels: Implications for the operation of doubly uniparental inheritance of mitochondrial DNA. Genome 41:818–824

Ghiselli F, Milani L, Passamonti M (2011) Strict sex-specific mtDNA segregation in the germ line of the DUI species Venerupis philippinarum (Bivalvia: Veneridae). Mol Biol Evol 28:949–961

Ghiselli F, Milani L, Chang PL, Hedgecock D, Davis JP, Nuzhdin SV, Passamonti M (2012) De novo assembly of the manila clam *Ruditapes philippinarum* transcriptome provides new insights into expression bias, mitochondrial doubly uniparental inheritance and sex determination. Mol Biol Evol 29:771–786

Gusman A, Lecomte S, Stewart DT, Passamonti M, Breton S (2016) Pursuing the quest for better understanding the taxonomic distribution of the system of doubly uniparental inheritance of mtDNA. PeerJ 4:e2760

Hayashi JI, Yonekawa H, Gotoh O, Watanabe J, Tagashira Y (1978) Strictly maternal inheritance of rat mitochondrial DNA. Biochem Biophys Res Commun 83:1032–1038

Hoeh WR, Blakey KH, Brown WM (1991) Heteroplasmy suggests limited biparental inheritance in Mytilus mitochondrial-DNA. Science 251:1488–1490

Hoeh WR, Stewart DT, Sutherland BW, Zouros E (1996) Cytochrome c oxidase sequence comparisons suggest an unusually high rate of mitochondrial DNA evolution in Mytilus (Mollusca: Bivalvia). Mol Biol Evol 13:418–421

Hutchison CA, Newbold JE, Potter SS, Edgell MH (1974) Maternal inheritance of mammalian mitochondrial DNA. Nature 251:536–538

Kenchington EL, MacDonald B, Cao L, Tsagkarakis D, Zouros E (2002) Genetics of mother-dependent sex ratio in blue mussels (Mytilus spp.) and implications for doubly uniparental inheritance of mitochondrial DNA. Genetics 161:1579–1588

Kenchington EL, Hamilton L, Cogswell A, Zouros E (2009) Paternal mtDNA and maleness are co-inherited but not causally linked in mytilid Mussels. PLoS One 4:e6976

Kiyomoto M, Komaru A, Scarpa J, Wada KT, Danton E et al (1996) Abnormal gametogenesis, male dominant sex ratio, and Sertoli cell morphology in induced triploid mussels, *Mytilus galloprovincialis*. Zoolog Sci 13:393–402

Kyriakou E, Zouros E, Rodakis GC (2010) The atypical presence of the paternal mitochondrial DNA in somatic tissues of male and female individuals of the blue mussel species *Mytilus galloprovincialis*. BMC Res Notes 3:222

Kyriakou E, Kravariti L, Zouros E, Rodakis GC (2013) Sex-specific mtDNA-protein interactions in a system of obligatory biparental inheritance and the exceptional role of perinuclear mitochondria. SW03S14-32

Kyriakou E, Chatzoglou E, Zouros E, Rodakis GC (2014a) The rRNA and tRNA transcripts of maternally and paternally inherited mitochondrial DNAs of Mytilus galloprovincialis suggest presence of a "degradosome" in mussel mitochondria and necessitate the re-annotation of the l-rRNA/CR boundary. Gene 540:78–85

Kyriakou E, Chatzoglou E, Rodakis GC, Zouros E (2014b) Does the ORF in the control region of Mytilus mtDNA code for a protein product? Gene 546:448–450

Kyriakou E, Kravariti L, Vasilopoulos T, Zouros E, Rodakis GC (2015) A protein binding site in the M mitochondrial genome of Mytilus galloprovincialis may be responsible for its paternal transmission. Gene 562:83–94

Ladoukakis MD, Zouros E (2001) Direct evidence for homologous recombination in mussel (*Mytilus galloprovincialis*) mitochondrial DNA. Mol Biol Evol 18:1168–1175

Ladoukakis M, Saavedra C, Magoulas A, Zouros E (2002) Mitochondrial DNA variation in a species with two mitochondrial genomes: the case of Mytilus galloprovincialis from the Atlantic, the Mediterranean and the Black Sea. Mol Ecol 11:755–769

Longo FJ, Dornfeld EJ (1967) The fine structure of differentiation of spermatid differentiation in the mussel, *Mytilus edulis*. J Ultrastruct Res 20:462–480

Machordom A, Araujo R, Toledo C, Zouros E, Ladoukakis EM (2014) Female-dependent transmission of paternal mtDNA is a shared feature of bivalve species with doubly uniparental inheritance (DUI) of mitochondrial DNA. J Zoolog Syst Evol Res. https://doi.org/10.1111/jzs.12096

Obata M, Sano N, Kawamura K, Komaru A (2007) Inheritance of two M type mitochondrial DNA from sperm and unfertilized eggs to offspring in *Mytilus galloprovincialis*. Dev Growth Differ 49:335–344

Obata M, Sano N, Komaru A (2011) Different transcriptional ratios of male and female transmitted mitochondrial DNA and tissue-specific expression patterns in the blue mussel, *Mytilus galloprovincialis*. Dev Growth Differ 53(7):878–886. https://doi.org/10.1111/j.1440-169X.2011.01294.x

Passamonti M, Ghiselli F (2009) Doubly uniparental inheritance: two mitochondrial genomes, one precious model for organelle DNA inheritance and evolution. DNA Cell Biol 28:79–89

Piganeau G, Gardner M, Eyre-Walker A (2004) A broad survey of recombination in animal mitochondria. Mol Biol Evol 21:2319–2325

Sano N, Obata M, Komaru A (2007) Quantitation of the male and female types of mitochondrial DNA in a blue mussel, *Mytilus galloprovincialis*, using real-time polymerase chain reaction assay. Dev Growth Differ 49:67–72

Sano N, Obata M, Komaru A (2010) Mitochondrial DNA transmitted from sperm in the blue mussel *Mytilus galloprovincialis* showing doubly uniparental inheritance of mitochondria, quantified by real-time PCR. Zoolog Sci 27:611–614

Skibinski DOF, Gallagher C, Beynon CM (1994a) Mitochondrial DNA inheritance. Nature 368:817

Skibinski DOF, Gallagher C, Beynon CM (1994b) Sex-limited mitochondrial DNA transmission in the marine mussel *Mytilus edulis*. Genetics 138:801–809

Stewart DT, Kenchington ER, Singh RK, Zouros E (1996) Degree of selective constraint as an explanation of the differentrates of evolution of gender-specific mitochondrial DNA lineages in the mussel Mytilus. Genetics 143:1349–1357

Sutovsky P, Moreno RD, Ramalho-Santos J et al (1999) Development – ubiquitin tag for sperm mitochondria. Nature 402:371–372

Sutovsky P, Moreno RD, Ramalho-Santos J et al (2000) Ubiquitinated sperm mitochondria, selective proteolysis, and the regulation of mitochondrial inheritance in mammalian embryos. Biol Reprod 63:582–590

Theologidis I (2007) The reversal of the transmission route of mitochondrial DNA in bivalves and its phylogenetic implications. PhD thesis, Department of Biology, University of Crete
Theologidis I, Fodelianakis S, Gaspar MB, Zouros E (2008) Doubly uniparental inheritance (DUI) of mitochondrial DNA in *Donax trunculus* (Bivanvia: Donacidae) and the problem of its sporadic detection in Bivalvia. Evolution 62:959–970
Tsaousis AD, Martin DP, Ladoukakis ED, Posada D, Zouros E (2005) Widespread recombination in published animal mtDNA sequences. Mol Biol Evol 22:925–933
Venetis C, Theologidis I, Zouros E, Rodakis GC (2006) No evidence for presence of maternal mitochondrial DNA in the sperm of *Mytilus galloprovincialis* males. Proc R Soc B 273:2483–2489
Venetis C, Theologidis I, Zouros E, Rodakis GC (2007) A mitochondrial genome with a reversed transmission route in the Mediterranean mussel *Mytilus galloprovincialis*. Gene 406:79–90
Verdonk NH, van den Biggelaar AM (1983) Earrly development and the formation of the germ leyers. In: Verdonk NH, van den JAM B, Tompa AS (eds) The Mollusca, vol 3. Development. Acamedic Press, New York, pp 91–122
White DJ, Wolff JN, Pierson M, Gemmell NJ (2008) Revealing the hidden complexities of mtDNA inheritance. Mol Ecol 17:4925–4942
Zouros E (2000) The exceptional mitochondrial DNA system of the mussel family Mytilidae. Genes Genet Syst 75:313–318
Zouros E (2013) Biparental inheritance through uniparental transmissions: the doubly uniparental Inheritance (DUI) of mitochondrial DNA. Evol Biol 40:1–31
Zouros E, Freeman RK, Oberhauser Ball A, Pogson GH (1992a) Direct evidence of extensive paternal mitochondrial DNA inheritance in the marine mussel Mytilus. Nature 359:412–414
Zouros E, Freeman RK, Oberhauser Ball A, Pogson GH (1992b) Direct evidence of extensive paternal mitochondrial DNA inheritance in the marine mussel Mytilus. Nature 359:412–414
Zouros E, Oberhouser Ball A, Saavedra C, Freeman KR (1994a) Mitochondrial DNA inheritance. Nature 368:817
Zouros E, Oberhauser Ball A, Saavedra C, Freeman KR (1994b) An unusul type of mitochondrial DNA inheritance in the blue mussel Mytilus. Proc Natl Acad Sci U S A 91:7463–7467

Exogenous Factors May Differentially Influence the Selective Costs of mtDNA Mutations

Wen C. Aw, Michael R. Garvin, and J. William O. Ballard

Contents

Abstract In this review, we provide evidence to suggest that the cost of specific mtDNA mutations can be influenced by exogenous factors. We focus on macronutrient-mitochondrial DNA interactions as factors that may differentially influence the consequences of a change as mitochondria must be flexible in its utilization of dietary proteins, carbohydrates, and fats. To understand this fundamental dynamic, we briefly discuss the energy processing pathways in mitochondria. Next, we explore the mitochondrial functions that are initiated during energy deficiency or when cells encounter cellular stress. We consider the anterograde response (nuclear control of mitochondrial function) and the retrograde response (nuclear changes in response to mitochondrial signaling) and how this mito-nuclear

W. C. Aw and J. W. O. Ballard (✉)
School of Biotechnology and Biomolecular Sciences, The University of New South Wales, Sydney, NSW, Australia
e-mail: z3314717@unsw.edu.au; w.ballard@unsw.edu.au

M. R. Garvin
School of Biological Sciences, Washington State University, Pullman, WA, USA
e-mail: michael.garvin@wsu.edu

P. Sutovsky (ed.), *Cellular and Molecular Basis of Mitochondrial Inheritance*, Advances in Anatomy, Embryology and Cell Biology 231, https://doi.org/10.1007/102_2018_2, Published online: 23 November 2018

crosstalk may be influenced by exogenous factors such as temperature and diet. Finally, we employ Complex I of the mitochondrial electron transport system as a case study and discuss the potential role of the dietary macronutrient ratio as a strong selective force that may shape the frequencies of mitotypes in populations and species. We conclude that this underexplored field likely has implications in the fundamental disciplines of evolutionary biology and quantitative genetics and the more biomedical fields of nutrigenomics and pharmacogenomics.

Keywords Fitness, Functional compensation, Metabolic flexibility

1 Introduction

Historically it was assumed that the consequences of a specific mutation were independent of the structural, cellular, metabolic, and physiological environment in which it occurred. This assumption is no longer valid. In the case of mitochondrial DNA (mtDNA), it has been established that the costs of specific mutations may be influenced by the sex of the organism in which it occurs and by its interaction with the nuclear-encoded subunits. Due to the maternal mode of inheritance of mitochondria in most metazoans, it has been argued that deleterious mutations that incur high costs to males could accumulate in the mtDNA if they are effectively neutral in females. These processes have collectively been termed the "mother's curse" effect (Ballard and Pichaud 2014; Connallon et al. 2018; Gemmell et al. 2004; Innocenti et al. 2011; Milot et al. 2017) and have recently been shown to contribute to the maintenance of a deleterious mitochondrial mutation in humans (Milot et al. 2017). Considering the complex interactions between the organelle and nuclear genomes, it has been argued that variations in the nuclear genome may compensate for slightly deleterious mtDNA changes to restore function (Chou and Leu 2010, 2015; Rand et al. 2004). In eukaryotes, ATP production by the mitochondrial electron transport system (ETS) depends on the interactions between the mitochondrial and nuclear genomes. The ETS includes five multi-subunit complexes (I–V), four of which are encoded by mitochondrial and nuclear genomes (McKenzie et al. 2007). In this review, we suggest it is also possible for exogenous factors such as diet to influence the homeostatic and organismal consequences of a nonsynonymous mtDNA change such that a mutation is detrimental in one context but beneficial in another. If true, a portion of the substantial variation in mtDNA sequences seen within populations could be functional and not simply due to a high rate of random mutation.

Mitochondrial functions are determined by the oxidation of dietary macronutrients, mechanistic interactions between the mitochondrial and nuclear genomes, the metabolic state of cells, and the life history of organisms. To understand these fundamental interactions, we start the review by briefly discussing the energy processing pathways in mitochondria. We then consider mitochondrial functions and explore the mitochondrial tasks that are initiated during energy deficiency or when cells encounter cellular stress. Finally, we focus upon mitochondrial Complex

I and discuss the potential role of diet as a strong selective force that may shape the frequencies of mitotypes in populations and drive speciation in primates. We conclude that exogenous factors, such as diet, may differentially influence the selective costs of mtDNA mutations and further studies of the factors influencing mitochondrial homeostasis are likely to give valuable insight into evolutionary biology and quantitative genetics as well as nutrigenomics and pharmacogenomics.

2 Mitochondrial Respiration

The oxidation of macronutrients occurs through the three different energy-producing pathways of glycolysis (cytosolic process), beta-oxidation of fatty acids (mitochondrial process), and oxidation of amino acids (cytosolic and mitochondrial process). Each of these pathways generates acetyl-CoA, which then reacts with oxaloacetate to enter into the mitochondrial tricarboxylic acid (TCA) cycle by forming citrate (Galluzzi and Kroemer 2014) (Fig. 1). In the TCA cycle, nicotinamide adenine dinucleotide (NADH) and flavin adenine dinucleotide ($FADH_2$) are produced as substrates to donate electrons to the ETS. Overall, mitochondrial complexes will collect electrons from the oxidation of substrates to reduce O_2 into H_2O. The "downhill" electron flow from Complex I to Complex V (ATP synthase) is coupled with an "uphill" proton transport. Protons are pumped from the mitochondrial matrix to the intermembrane space to form a transmembrane electrochemical gradient. The transmembrane flow of protons down the electrochemical gradient through Complex V is coupled with the phosphorylation of adenosine diphosphate (ADP) to adenosine triphosphate (ATP) (Nelson et al. 2008). Thus, mitochondria convert metabolites from macronutrient catabolism into usable energy ATP. Following the second law of thermodynamics (Atkins 1994) when free energy is converted into usable energy (ATP) in mitochondria, some energy is lost in the form of heat and reactive oxygen species (ROS) (Fig. 1).

During the process of converting free energy into usable energy, metabolites will be consumed and produced. These mitochondrial metabolites include ATP, ADP, AMP, nicotinamide adenine dinucleotide (NAD), and ROS. ATP is hydrolyzed to ADP by the reaction $ATP + H_2O \rightarrow ADP + P_i$ + free energy: with the calculated free energy for the hydrolysis of 1 mol of ATP being −57 kJ/mol (Robergs et al. 2004). Likewise, energy is released when phosphate is removed from ADP to form AMP. The free energy can be transferred to other molecules and make less favorable reactions in a cell more favorable. AMP can then be recycled into ADP or into ATP by forming new phosphoanhydride bonds to store energy. In the cell, AMP, ADP, and ATP are interconverted continuously as they participate in biological reactions. NAD is a coenzyme found in all living cells. It exists in two forms: an oxidized form (NAD^+) and a reduced form (NADH). NADH takes part in redox

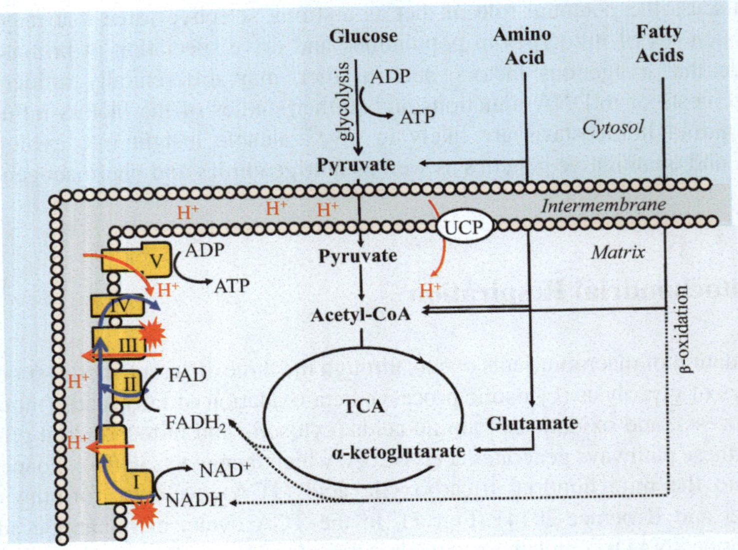

Fig. 1 Macronutrient oxidation. Oxidation of macronutrient via glycolysis, beta-oxidation, and protein metabolism provides compounds that feed into the tricarboxylic acid (TCA) cycle. The TCA cycle generates substrates (NADH and $FADH_2$) for oxidative phosphorylation into Complexes I and II, respectively. Mitochondrial complexes will collect electrons (purple line) from substrate oxidation and transport protons (red line) into the intermembrane space. The transmembrane flow of protons (H^+) down the electrochemical gradient through Complex V is coupled with phosphorylation of ADP to ATP. During the process of oxidative phosphorylation, reactive oxygen species (red explosive symbol) are generated as a by-product. Cellular energy can also be obtained through glycolysis of glucose. The uncoupler proteins (UCP) provide an alternative route for proton to reenter the mitochondrial matrix. I, II, III, IV, and V are Complex I, Complex II, Complex III, Complex IV, and Complex V, respectively

reactions and is responsible for transferring electrons from one molecule to another. Both NAD^+ and NADH are crucial in biological processes including energy metabolism (Ying 2008), calcium homeostasis (Ziegler 2000), gene expression (Gomes et al. 2013; Rutter et al. 2001), and aging (Gomes et al. 2013). More than 90% of total cellular ROS is produced during the process of mitochondrial respiration (Andreyev et al. 2005), and seven major sites of superoxide production are known in mammalian mitochondria (Brand 2010). Of these, Complexes I and III in ETS are the primary producers of ROS (Fig. 2) (Sugioka et al. 1988; Turrens et al. 1985). Low levels of ROS may have a benefit through mitochondrial hormesis (mitohormesis) (Ristow and Schmeisser 2014), whereby ROS act as signaling molecules that provide a fitness advantage through higher antioxidant capacity (Ristow and Schmeisser 2014; Tapia 2006). However, elevated ROS levels are predicted to reduce longevity primarily through DNA damage and also through damage to proteins and fats (Sohal and Weindruch 1996).

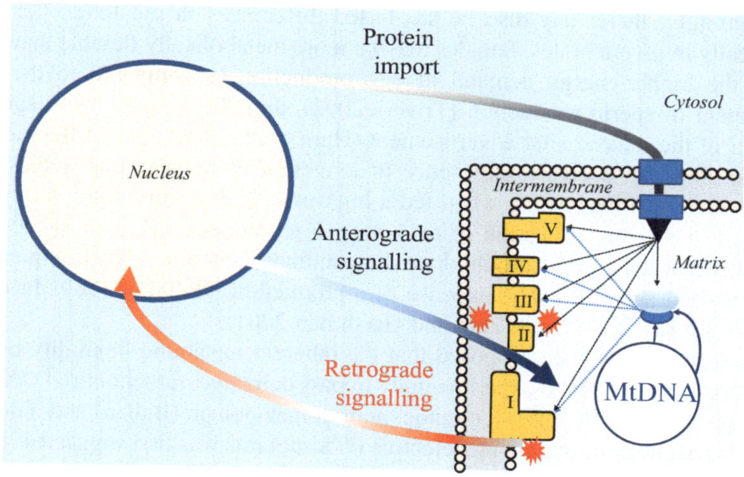

Fig. 2 Communication between the nucleus and mitochondrion. Proteins imported into the mitochondrion are essential for building all complexes of the electron transport chain, while proteins encoded by the mtDNA are essential for Complexes I, III, IV, and V. Nuclear and mitochondrial proteins are crucial for the mitochondrial ribosome (blue ellipses). Anterograde response functions primarily to regulate mitochondrial gene expression. Retrograde signaling informs the nuclear genome of the health of the organelle. During the process of oxidative phosphorylation, reactive oxygen species (red explosive symbol) are generated as a by-product. These species appear to be a key trigger for the mitochondrial retrograde response. I, II, III, IV, and V are Complex I, Complex II, Complex III, Complex IV, and Complex V, respectively

2.1 Metabolic Flexibility

By necessity, the mitochondrion is a metabolically flexible organelle. The organelle can adapt fuel oxidation (glycolysis, amino acid oxidation, and beta-oxidation) to fuel availability (carbohydrate, protein, and fat). Metabolic flexibility allows the organism to adapt to changes in the dietary macronutrient ratio (van de Weijer et al. 2013) and alters the production of mitochondrial metabolites including ATP and ROS. For example, in optimal conditions, complete oxidation of a molecule of glucose by ETS in mitochondria will produce approximately 15 times more ATP than glycolysis (Alberts et al. 2002). Furthermore, ATP generated independent of mitochondria through glycolysis does not cause an increase in ROS production, whereas ATP made from fatty acids and amino acids depends on mitochondrial oxidative phosphorylation (OXPHOS) and produces ROS as a by-product (Fig. 1) (Ristow and Schmeisser 2014). Therefore, for the same ATP yield, carbohydrate metabolism will result in less ROS than protein metabolism. Mutations in mtDNA have also been shown to cause a shift from OXPHOS respiration to glycolysis (Jayasena et al. 2015; Morris and Berk 2015; O'Donnell and Ahuja 2005) and lead to lipid accumulation (Lee et al. 2013). Thus, changes in energy metabolism caused by either mtDNA mutations or dietary changes can affect physiological traits that require high energy input or high lipid demands.

Intriguingly, there may also be sex-based differences in metabolic flexibility. Specifically in invertebrates, females may be more metabolically flexible than males due to the higher-energy demand of egg production as compared to the lower requirement of sperm production (Trivers 2002), though this may also depend on the level of the male sexual advertisement (Hunt et al. 2004). This difference may then translate into a greater difference in longevity of females than males under extreme diets. All else equal, when fed a high-protein diet, survival is likely to be shortened due to the allocation of resources to reproduction (Lee et al. 2008). In contrast, when fed a high-carbohydrate diet, animals switch to a physiological state geared toward survival at the expense of reproduction (Halliday 1989; Jacob and Evans 2000; Lee et al. 2008; Zera and Harshman 2001).

In humans, it has been proposed that the inherent metabolic flexibility of mitochondria can be utilized to experimentally bypass defective mitochondrial OXPHOS complexes and thereby influence mitochondrial metabolism (Ballard and Youngson 2015). The strategy of bypassing defective ETS subunits was first suggested by Eleff et al. (1984), who demonstrated that supplementation could maintain muscle ATP synthesis in a patient with a genetically defective form of Complex III with ascorbate (vitamin C) and menadione (vitamin K3). Subsequently, it was shown that infusion of succinate, which bypasses Complex I, prevented decline in liver ATP levels and prolonged survival in cases of experimental sepsis, a pathology involving energy dysfunction (Ferreira et al. 2000; Protti and Singer 2006). In cultured fibroblasts from a patient with Leigh syndrome due to a genetic mutation in Complex I, succinate prodrugs added in vitro to the cells improved mitochondrial respiration (Ehinger et al. 2016). However, this strategy of bypassing defective ETS subunits has critical bioenergetic consequences. Differences in proton concentration caused by different entry points into the ETS will influence the transmembrane electrochemical gradient and eventually affect the rate of ATP produced by Complex V (Ballard and Youngson 2015). For example, electrons entering through Complex I can generate a total of 10 H^+. However, only 6 H^+ can be produced when the electrons bypass Complex I and enter through Complex II. In the next section, we explore the mitochondrial functions that are involved in metabolic compensation.

3 MtDNA and Mitochondrial Functions

The endosymbiotic theory hypothesizes that a mutualistic relationship was formed when a eukaryotic cell with glycolytic energy production merged with an α-proteobacteria that produces energy via OXPHOS (Sagan 1967). Over time, this relationship has evolved by transferring most of the mitochondrial genome to the nuclear genome. Currently, the nuclear genome encodes proteins required for glycolysis and approximately 90 proteins that make up the ETC (Fig. 2). In contrast, the mitochondrial genome retains most of the essential genes required to initiate protein translation, but its functions are dependent on the nucleus genome to provide enzymes needed for DNA replication, repair, transcription, and translation (Wallace 2009; Whelan and Zuckerbraun 2013).

In healthy individuals, continuous cellular signaling fine-tunes the interactions between the mitochondrial and nuclear genomes. In this section, we first consider mito-nuclear crosstalk involving the anterograde response, which is defined as nuclear control of mitochondrial function. The anterograde response is essential for mtDNA-encoded protein synthesis as this process depends entirely on the nucleus for supplying the enzymes that are required for replication, repair, transcription, and translation (Almeida et al. 2012). We then consider the retrograde response, which involves nuclear changes that respond to mitochondrial signaling. The retrograde response is essential for reconfiguration of metabolism that may include compensation of mitochondrial defects. The final segment of this section considers how the tight coordination between anterograde and retrograde signaling is influenced by exogenous factors such as diet and temperature.

3.1 Anterograde Response

Mitochondria are heavily dependent on the nuclear genome to contribute much of their functionality. The anterograde response which functions to regulate mitochondrial gene expression can be initiated by environmental and endogenous stimuli such as stress, nutrient availability, and physical factors (Thiriet 2011). These anterograde responses include AMP-activated protein kinase (AMPK) pathway, cAMP-protein kinase A pathway, and calcium signaling (Quiros et al. 2016). Here, we use the AMP-activated protein kinase (AMPK) pathway to explain the anterograde signaling events that involve the activation of the transcription cascade. AMPK is the primary regulator of mitochondrial biogenesis and acts as a biological sensor that is activated by changes in AMP/ATP ratio. Reduction in cellular energy caused by metabolic stress (e.g., decrease in cellular glucose level or increase in physical activity) will increase the AMP/ATP ratio and activate the AMPK system (Corton et al. 1994; Fujii et al. 2000; Wojtaszewski et al. 2000). The activation of AMPK, in turn, increases the cellular NAD^+ level and lead to activation of sirtuin 1 (SIRT1). SIRT1 is responsible for the regulation of the PPARγ coactivator 1α (PGC1α) which functions to control mitochondrial energy metabolism and biogenesis (Quiros et al. 2016). One role of PGC1α is to control mitochondrial gene expression by increasing nuclear respiration factor 1 (NRF-1) binding activity (Jornayvaz and Shulman 2010). This will induce an increase in transcription of multiple respiratory genes, including COX, ATP synthase, mitochondrial transcription factor-A, mtDNA polymerase gamma, as well as genes encoding proteins that are responsible for mitochondrial protein import (Bachman et al. 1996; Evans and Scarpulla 1990; Scarpulla 2008; Suliman et al. 2003).

Other than controlling mitochondrial gene expression, there is accumulating evidence suggesting that AMPK also functions to regulate cellular growth and autophagy (Mihaylova and Shaw 2011). When ATP is depleted, AMPK inhibits cell growth and proliferation by switching off the target of rapamycin (TOR), a downstream target of AMPK that regulates mitochondrial OXPHOS activity and

ROS homeostasis (Colombani et al. 2003; Jacinto and Hall 2003). Autophagy is a cellular process of "self-engulfment" where the cell breaks down its organelles (macroautophagy) and cytosolic components (microautophagy) to ensure sufficient metabolites when nutrients run low (Mihaylova and Shaw 2011). In the extreme case of starvation, the breakdown of cellular components promotes cellular survival by maintaining cellular energy levels. In disease, autophagy has been seen as a response to stress. Next, we consider the retrograde response, which includes numerous mechanisms that the mitochondria uses to regulate nuclear gene expression.

3.2 *Retrograde Response*

The retrograde response is well understood in yeast and *Caenorhabditis elegans* but less comprehended in *Drosophila* and mammals. Most generally, the response can be initiated by mitochondrial metabolites such as mitochondrial ROS, NAD^+, NADH, and Ca^{2+} (Butow and Avadhani 2004; Jazwinski 2013) and may function, at least partly, as a compensatory mechanism for less efficient mitochondria (Lee and Wei 2000). An increase in mitochondrial ROS production induced by mitochondrial dysfunction may elicit chronic oxidative stress, which in turn has been shown to increase mtDNA copy number and mitochondrial gene expression (Passos et al. 2007; Yoboue and Devin 2012). Furthermore, empirical studies have reported that mitochondrial ROS may mediate control of mitochondrial biogenesis through crosstalk between the nucleus and mitochondria (Yoboue and Devin 2012). Mitochondrial biogenesis is believed to be controlled by mitochondrial ROS via regulation of the transcription factor NRF-1 (Yoboue and Devin 2012). NRF-1 expression has been shown to increase with the addition of hydrogen peroxide to lung fibroblasts (Lee et al. 2002), while both NRF-1 and NRF-2 expression is upregulated by lipopolysaccharide-induced oxidative stress damage (Suliman et al. 2003). Both NRF-1 and NRF-2 regulate the transcription of multiple respiratory genes involved in controlling mitochondrial functions (Bachman et al. 1996; Evans and Scarpulla 1990; Scarpulla 2008; Suliman et al. 2003).

To develop a greater understanding of the retrograde response in humans, Chae et al. (2013) genetically modified cybrid cells to have varying amounts of mtDNA with the mt3243 mutation. An A to G substitution at base pair 3243 in the dihydrouridine loop of mitochondrial $tRNA^{Leu}$ is commonly associated with muscular dysfunction, maternally inherited diabetes, and deafness (Goto et al. 1990). By analyzing the gene expression profiles in cells with the mutant $tRNA^{Leu}$ and the transcription factors that recognize the differentially regulated genes, Chae et al. (2013) identified 72 transcription factors associated with the retrograde response and experimentally validated retrograde signaling pathways involving ROS, PGC1α, RXRA (retinoid X receptor α), and kinase JNK (c-JUN N-terminal kinase). Such studies increasing our understanding of retrograde pathways will help us determine the influence a single mtDNA mutation may have on eliciting a response by the nuclear genome and may help develop novel treatments for mitochondrial disease. In some cases, modifying the nuclear genome to compensate for the mtDNA mutation may help to restore function. Next, we consider diet and temperature as exogenous factors that may influence mitochondrial homeostasis.

3.3 Exogenous Factors

Diet and temperature may influence the evolution of mitochondria and mtDNA. An interesting study in bats revealed that positive selection is acting on the enzyme alanine-glyoxylate aminotransferase 1 (AGT) in a diet-dependent manner (Liu et al. 2012). The AGT enzyme has a vital role in detoxifying glyoxylate, an intermediate metabolite that would otherwise be converted to the harmful oxalate (Ichiyama 2011). In carnivorous mammals, the precursor of glyoxylate is hydroxyproline, which occurs in the mitochondria, whereas in herbivores, the primary precursor is glycolate, which accumulates in peroxisomes (Noguchi 1987; Takayama et al. 2003). Employing bats as a model, Liu et al. (2012) found that all species preying on invertebrates or vertebrates (high protein) have retained the mitochondrial targeting sequence (MTS), whereas multiple herbivorous bat species that feed on fruits (high carbohydrate) have independently lost MTS. Firmly, these studies show that diet strongly influences mitochondrial evolution.

Dietary changes may also drive population subdivision in mtDNA. The first evidence showing that diet can modulate mitochondrial evolution came from a phylogenetic analysis of 277 human mitochondrial genomes (Kivisild et al. 2006). This study assessed individuals' mtDNA from five continents and found that the direction of amino acid substitutions varied among populations. Human mitochondrial haplogroup H1 from Europe and the Near East was found to have more nonsynonymous mtDNA mutations that removed threonine and valine, while haplogroup V from Finish populations has more substitutions to threonine or valine. Interestingly, Finland's population greatly relies on hunting and fishing, and they are known to consume a high-protein diet, where both threonine and valine are found abundantly in meat or fish but are deficient in most grains (Kivisild et al. 2006). A limitation of this study is that the relative proportions of threonine and valine in the nuclear genome are not reported. More recently, Towarnicki and Ballard (2017) assayed the larval development time of two *Drosophila* mitotypes (Alstonville and Dahomey) in two nuclear genetic backgrounds. In both nuclear genetic backgrounds, larvae with Dahomey mtDNA developed more slowly than Alstonville when they were fed a high-protein diet at all temperatures. However, when fed a high-carbohydrate diet, Dahomey larvae matured more quickly than Alstonville at 23 and 27 °C indicating an evolutionary advantage. These authors proposed that the V161L mutation in mtDNA-encoded NADH dehydrogenase subunit 4 of Complex I was likely driving these startling results.

Regional variations in mtDNA haplotypes have been argued to be a consequence of adaptation to colder climate (Ballard and James 2004; James and Ballard 2003; Mishmar et al. 2003; Ruiz-Pesini et al. 2004), which was likely necessary for the ancestors of modern humans to migrate out of Africa (Ruiz-Pesini et al. 2004; Stringer and Andrews 1988). This is analogous to Bergmann's rule that species and populations of larger size are found in colder climates as an adaptive mechanism. As mitochondria play an essential role in ATP production and body temperature maintenance, the balance between ATP production and heat generation is believed to be a fundamental explanation for environmental adaptation. In cold climates, mitochondrial haplotypes with lower coupling efficiency may be expected to be more beneficial as a reduction in coupling efficiency will increase thermogenesis and therefore increase cold resistance (Ruiz-Pesini et al. 2004). However, heat production from less coupled mitochondria would necessitate a high caloric diet or increased food intake. In less coupled mitochondria, the presence of excess calories and ADP would cause constant transfer of electrons and H^+ into the ETS and mitochondrial intermembrane space, respectively (Michelakis 2008). The less coupled mitochondria will "short circuit" H^+ out into the mitochondrial matrix either through UCP or "leaky mitochondrial complexes" that are caused by slightly deleterious mtDNA mutations. This can be beneficial as the leakage of H^+ from UCP or "leaky mitochondrial complexes" can result in mitochondrial thermogenesis which could allow better adjustment to cold climate (Michelakis 2008).

If this hypothesis is true, mutations that result in less coupled mitochondria would require a higher caloric intake to fuel the energy losses from the generation of heat, and the organisms would be predicted to have a higher basal metabolic rate (BMR). This postulation is supported by data from human studies showing that circumpolar populations have a tendency of consuming high-fat diets (Fumagalli et al. 2015). In addition, evidence also shows that the BMR of circumpolar populations is higher than temperate populations and these differences in BMR are associated with UCP gene variations (Bhopal and Rafnsson 2009; Leonard et al. 2002; Ruiz-Pesini et al. 2004). UCP is an important protein that regulates thermogenesis and fatty acid metabolism (Argyropoulos and Harper 2002). Further, a high-fat diet not only provides high calories for heat generation but also functions to stimulate the upregulation of UCP to increase mitochondrial uncoupling and fatty acid metabolism (Cole et al. 2011).

Alternatively, the adaptation to colder climates could be related to food availability with temperature being a correlated variable (food is scarce in cold months). In fact, Bergmann's rule mentioned previously has been suggested to be invalid, and body size best predicted by the duration of the annual productivity pulse (Geist 1987). Food availability was proposed as an explanation for the discovery of a strong signature of selection at the ND5 subunit of mitochondrial Complex I (Garvin et al. 2015), but what was missing at the time was a means to accurately predict the functional implications of those selected sites. In the next section, we focus upon Complex I mutations to study dietary-mitochondrial interactions. Mitochondrial Complex I (NADH: ubiquinone oxidoreductase) is the first enzyme of the ETS and is composed of 7 mtDNA-encoded and 38 nuclear-encoded subunits in humans. These subunits must work synergistically to function optimally.

4 Mitochondrial Complex I Case Study

Complex I exists in a supercomplex with respiratory chain Complexes III and IV and is required for the generation of a transmembrane proton gradient used for ATP synthesis (Fig. 3). Complex I transfers two electrons from NADH to ubiquinone and pumps four protons across the inner mitochondrial membrane. It produces approximately 40% of the proton flux that is used for ATP production. Structures for all five mitochondrial complexes are now available for higher vertebrates, and significant advances have been made on how each of these function at the molecular level. Evolutionary studies can inform function because sites that are under positive selection should by definition change function to optimize for local adaptation. In Pacific salmon, it has been proposed that changes in the mtDNA-encoded proton-pumping piston arm of Complex I influence organismal fitness (Garvin et al. 2015). In humans, dysfunctions of the mitochondrial Complex I subunits are associated with a variety of phenotypes including reduced fertility, shortened life span, progressive neurodegeneration, insulin resistance, type 2 diabetes, an increase in ROS production, and a decrease in ATP production (Aw et al. 2011; Burman et al. 2014; Clancy 2008; Correa et al. 2012; Schon and Manfredi 2003; Selvi Rani et al. 2006; van de Weijer et al. 2013). Recent reviews have suggested the complex is under

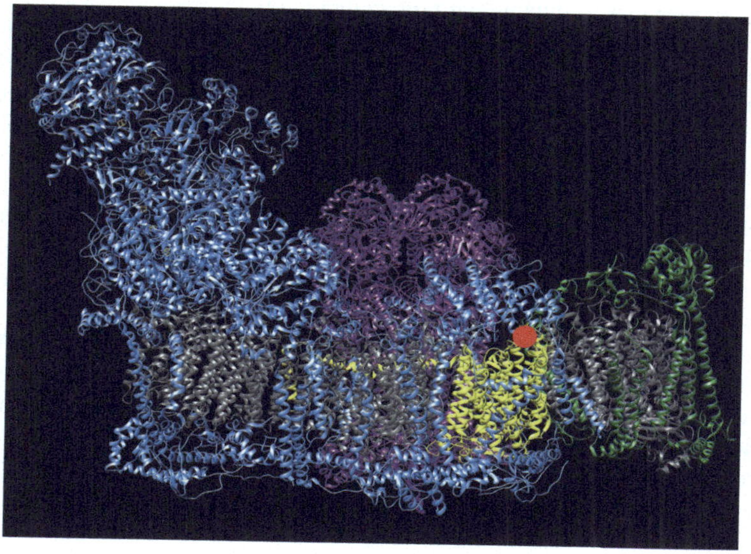

Fig. 3 Cryo-EM structure of human respiratory supercomplex I/III/IV. Gray residues are mitochondrial-encoded subunits. The ND5 subunit (yellow) displays a strong signature of positive Darwinian selection in numerous taxa including hominids. One site shown to be under selection is ND5 515 (red circle). The nucleotide positions that alter ND5 515 are 13,880 and 13,879 from NC_012920. One example of a selected amino acid change occurs in Gorilla (S515H). Modified from Guo et al. (2017)

strong selection in a wide variety of taxa (Ballard and Youngson 2015; Garvin et al. 2015) and an increasing number of studies have investigated the relationship between macronutrients and mitochondrial functions in organisms with Complex I dysfunction (Aw et al. 2017; Correa et al. 2012; Pichaud et al. 2013; Towarnicki and Ballard 2017).

In the next section we consider the influence of protein, carbohydrate, and fat on organisms and organelles that harbor mtDNA with Complex I mutations. We then consider Complex I mutations in primates. Based on modeling data, a specific mtDNA mutation can be predicted to be structurally significant. However, it is less clear whether the change will be functionally significant due to the metabolic flexibility of the mitochondrion and the ability for the organelle to compensate or overcompensate for the mutation on a given diet. Here we posit that a Complex I mutation has the potential to change the retrograde response by increasing ROS levels and reduce the bioenergetic reliance on the complex, increasing organismal fitness.

4.1 Influence of Dietary Protein on Complex I Mutations

We hypothesize that feeding a high-protein diet to organisms with slightly deleterious Complex I mutations will result in increased mitochondrial ROS production that activates a retrograde response. The retrograde response is expected to elicit a compensatory antioxidant reaction as well as a homeostatic anterograde response that upregulates mitochondrial biogenesis to meet the cellular energy demand (Jornayvaz and Shulman 2010). In some cases, a small increment of ROS production will be beneficial for the organism as it increases mitochondrial density, ATP production, and fecundity (Sena and Chandel 2012) (Fig. 4). In other cases, substantial increases of ROS levels are expected to be cytotoxic and cause cellular damage. In *Drosophila*, a Complex I ND2 mutation has been shown to increase mitochondrial ROS production and impair energy production (Correa et al. 2012; Pichaud et al. 2013). This ND2 mutation also induced an increase in mtDNA copy number but with a trade-off of reduced life span (Clancy 2008; Correa et al. 2012). If reproduction occurs early in an organism's life, or the animal is short-lived, then the evolutionary advantage of increased fecundity may outweigh the disadvantage of reduced longevity. Conversely, increasing ROS production may be deleterious for long-lived animals where reproduction is extended over an organism's life span. Therefore, there are complex links between dietary macronutrients, mtDNA mutations, the metabolic state of cells, and the life history of animals that require further study.

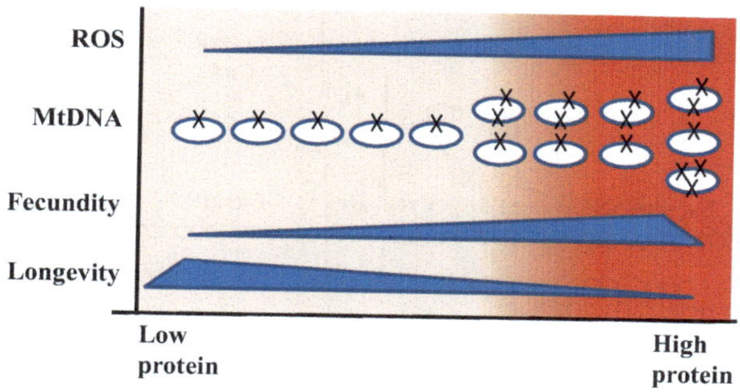

Fig. 4 The consequences of a mtDNA mutation are diet specific. An increase in dietary protein intake causes an upregulation of ROS. Low levels of mitochondrial ROS function as signaling molecules to maintain cellular homeostasis. In healthy individuals, increasing ROS levels cause an increase in mtDNA copy number (ellipse) and result in higher levels of ATP and increased fecundity, up to a point. In contrast, high levels of mitochondrial ROS induce cytotoxicity and cause damage to proteins, lipids, and nucleic acids. In animals with a slightly deleterious mtDNA mutation (black cross), increasing copy number can be detrimental as it feeds back to increase ROS production, which causes somatic mutations to accumulate (additional black crosses) and longevity is reduced

4.2 Influence of Carbohydrate on Complex I Mutations

We posit that an organism harboring a slightly deleterious Complex I variation may respond in one of two ways to a high carbohydrate diet. Plausibly the animal may switch energy production from a mitochondrial-dependent (OXPHOS) to a mitochondrial-independent (glycolysis) pathway (Fig. 5). Alternatively, there may be a biochemical bypass of Complex I to reduce its influence (Fig. 5). Conversion of glucose to lactic acid in the presence of oxygen is known as aerobic glycolysis or the "Warburg effect." Increased aerobic glycolysis is observed in cancer cells and was first reported by Warburg in the 1920s (Koppenol et al. 2011; Warburg et al. 1924), leading to the hypothesis that cancer results from impaired mitochondrial metabolism. Although the "Warburg hypothesis" has proven to be incorrect, the experimental observations of increased glycolysis in tumors even in the presence of oxygen have been repeatedly verified (Semenza et al. 2001). It remains unclear why cancer cells employ the energetically less efficient aerobic glycolysis, rather than oxidative phosphorylation, which generates far more ATP from a single glucose molecule. However, Calabrese et al. (2013) demonstrated that mitochondrial Complex I is required to perform the metabolic switch toward glycolysis during the hypoxia response, a crucial goal in the metabolic reprogramming of cancer cells. Their results strongly indicate that the mechanism through which Complex I regulates the glycolytic shift is mediated by HIF-1α stabilization, allowing tumor cells to adapt to hypoxia and persist in proliferation.

Mutations in nuclear-encoded genes may also influence mitochondrial homeostasis and switch to aerobic glycolysis. In *Drosophila* the nuclear-encoded mitoribosomal protein S12 (mRpS12) is encoded by the gene *technical knockout* (*tko*).

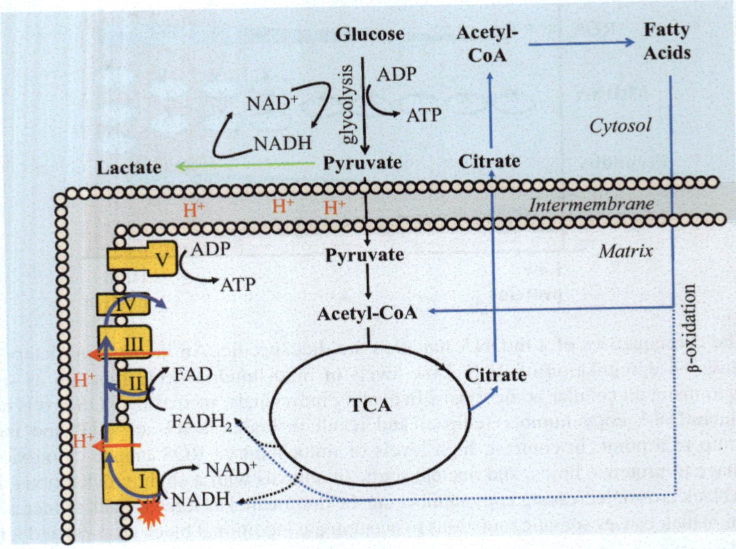

Fig. 5 Pyruvate can be formed in the cytosol by glycolysis. Mutations in mitochondrial Complex I may cause a shift from mitochondrial OXPHOS to glycolysis (green arrow). Hence, less ROS and more lactate are produced. Pyruvate imported into mitochondria can be oxidized into acetyl-CoA. Pyruvate oxidation is vital for the production of reducing equivalents for ATP synthesis, and citrate formed in the TCA cycle can be exported to the cytosol, converted to acetyl-CoA, and used to produce new fatty acids (blue arrow). These fatty acids can be oxidized through ß-oxidation to produce acetyl-CoA, NADH, and $FADH_2$, which partially bypass Complex I. Carboxylation is an alternative fate for mitochondrial pyruvate. In this case, oxaloacetate is converted to malate and exported to the cytosol

The gene name in *Drosophila* reflects the so-called bang-sensitive phenotype of the canonical allele, tko^{25t}, which suffers paralytic seizures induced by mechanical stress. Null alleles of *tko* are larval-lethal, but the tko^{25t} phenotype is relatively mild and thus constitutes an animal model for mitochondrial disorders. The mutation reduces Complex I activity and has been shown to alter the pattern of gene expression in *Drosophila* (Chen et al. 2012; Fernández-Ayala et al. 2010; Jacobs et al. 2004). Transcriptome-wide analysis of gene expression shows the tko^{25t} mutant flies experience a metabolism switch to compensate the OXPHOS defect. Specifically, the tko^{25t} mutant flies undergo an upregulation of lactate dehydrogenase which suggests a biochemical switch to glycolysis for ATP production (Fernández-Ayala et al. 2010).

Remodeling pyruvate metabolism can partially bypass Complex I and cause physiological changes including an increase in fat (Fig. 5). Once imported into the mitochondrial matrix, pyruvate can be metabolized by different routes (McCommis and Finck 2015). Most pyruvate in oxidative tissues is converted to acetyl-CoA. This acetyl-CoA then enters the TCA cycle, and these carbons are predominantly converted to CO_2. The energy created by turning pyruvate into CO_2 produces the reducing equivalents NADH and $FADH_2$, which are essential in generating the proton gradient required for OXPHOS and ATP production. Alternatively, acetyl-CoA can also be used in the anabolic production of fatty acids (Fig. 5). In this case, acetyl-CoA is combined with oxaloacetate by citrate synthase to form citrate. This citrate can exit the mitochondria and is cleaved back to acetyl-CoA and oxaloacetate in the cytosol (McCommis and Finck 2015). Fatty acids are then synthesized from acetyl-CoA by the process of lipogenesis. Three fatty acids may then combine with glycerol to form triglycerides, which are the main constituents of body fat in most animals including humans. When needed, lipolysis then breaks down the fats and other lipids by hydrolysis to release fatty acids. Short- and medium-chain fatty acids (up to 12 carbons in length) can freely enter the mitochondrial matrix, while long-chain fatty acids must be transported into the mitochondria by the carnitine shuttle. β-oxidation involves the sequential removal of two-carbon segments in the form of acetyl-CoA and production of shortened acyl-CoA, with concomitant reduction of one FAD and one NAD^+. The electrons carried by NADH and $FADH_2$ immediately enter the ETS, whereas acetyl-CoA enters the TCA cycle (Fig. 5). ß-oxidation of fatty acids produces more $FADH_2$ than the oxidation of amino acid or glycolysis.

4.3 Influence of Fat on Complex I Mutations

Differences in dietary fat intake may modulate mitochondrial functions through mitochondrial fission and fusion and changes in metabolic metabolism. Mitochondrial fission is responsible for removing damaged organelles through the process of autophagy. Mitochondrial fusion has been suggested to maximize the mitochondrial respiratory capacity by allowing efficient mixing of mitochondrial content (Westermann 2012). It has been shown that omega-3 fatty acid reduces ROS production by increasing the expression of mitofusin 2 (*Mfn2*), a protein that is essential for mitochondrial fusion (Lionetti et al. 2014; Zhang et al. 2011). In contrast, saturated fatty acid diets are well known to elicit hepatic lipid accumulation, which induces mitochondrial dysfunction (Hirabara et al. 2010) and fragmentation (Jheng et al. 2012), but the results remain controversial. Notably, saturated fatty acids with C18:0 chain lengths have been reported to boost the mitochondrial fusion in *Drosophila* (Senyilmaz et al. 2015).

In most wild animals, high-fat diets are rare except in cold climates. As suggested, a high-fat diet helps regulate temperature circumpolar populations. In more temperate regions, eating an excessive amount of fat causes many metabolic diseases such as type II diabetes, atherosclerosis, and cardiovascular diseases. Despite its incompletely known metabolic functions, the high-fat ketogenic diet is increasingly recommended for the treatment of some intractable epilepsies (Youngson et al. 2017). All of the variants of the ketogenic diet instigate shifts in energy metabolism toward fat-based energy generation and away from glucose-based. The ingested fatty acids are metabolized in liver mitochondria into ketone bodies which are then released into the bloodstream for uptake by other organs including the brain. In the mitochondria of neurons and glial cells, ketone bodies are catabolized to acetyl-CoA, which can then enter the TCA cycle for energy generation, or it can be used in lipogenesis to produce fatty acids. A consequence of the increased dependence on mitochondria for energy generation is that the number of organelles increases in neurons and glia which leads to high levels of ATP production (DeVivo et al. 1978; Rho 2017). As described, ß-oxidation of fatty acids provides a higher proportion of electron downstream of Complex I. Mutations in nuclear-encoded subunits of Complex I have been linked with epilepsy (Schuelke et al. 1999); therefore, decreasing reliance on Complex I may be an advantage for a subset of people with epilepsy.

4.4 Influence of Complex I Mutations in Primates

Advances in structural modeling enable new insights into the links between mtDNA mutations and concomitant changes in mitochondrial functions, life history traits, and disease. The recent structure of the mammalian mitochondrial supercomplex (Wu et al. 2016) now allows one to not only pinpoint the location of the sites under selection in the ND5 subunit of vertebrates (Garvin et al. 2015) but also the nuclear-encoded subunits of complexes I, III, and IV. Site 515 in the ND5 gene was shown to be under positive Darwinian selection during the evolution of Hominidae (Garvin et al. 2015), but the site did not appear to interact with residues in Complex I. The supercomplex structure, however, suggests that this site interacts with the nuclear-encoded COX7A subunit of Complex III (Fig. 3).

COX7A has two isoforms (A1 and A2) in humans that are expressed in different tissues. COX7A1 is found predominantly in the heart and skeletal muscle, whereas COX7A2 is considered the liver isoform. The liver isoform has an alanine in place of the threonine that appears to interact with the site at 515 of ND5 through hydrogen bonding and therefore is likely not affected by this substitution. However, the muscle and heart isoform, necessary for high oxygen consumptive tissues, is predicted to decrease coupling between Complexes I and III in cases where hydrogen bonding is lost (Asn, Pro) and increase coupling with basic residues (His). Substitutions from serine to threonine may cause subtler changes in the orientation and strength of the hydrogen bond. Other positively selected sites in the Helix HL of Complex I also appear to affect the coupling of the supercomplex, which suggests that the rate of energy flux through the supercomplexes is the functionally selected trait that may be linked to differences in food availability among environments and energetic requirements of the species that inhabit them.

Differences in dietary constraints support the theory that nutrition and food availability have the potential to drive speciation in higher primates (Yamagiwa and Basabose 2009) and may explain the relationship between brain development and bioenergetics. Species within *Homo* need high-quality foods throughout the year because 20–50% of their energetic resources are likely used by their brain, (Leonard and Robertson 1992, 1994; Ponce de Leon et al. 2008). It has been shown that fetal brain growth varies among hominids (Leigh 2004) and that this likely reflects two different life history strategies. The first involves rapid prenatal brain growth followed by a slowed postnatal brain growth rate combined with late reproductive maturation of mothers. These species require prolonged food availability to provide maternal investment through lactation. The second life history strategy is represented by species that mature early but have slower prenatal brain growth followed by more rapid postnatal brain development. These species need abundance of high-quality food for a shorter period than the species in the first group because of the shorter postnatal development period. The offspring of species with the second strategy must either provide food for themselves or receive it from members of the community (Leigh 2004; Ulijaszek 2002). In the latter case, mitochondrial Complex I variation is predicted to be correlated with the rate of brain development during weaning.

5 Conclusion

Current evidence suggests that diet can modulate mitochondrial functions and thereby affect the fitness of the organisms (Aw et al. 2018). Plausibly, different mitotypes will incur distinct fitness costs under different macronutrient ratios. Thus, geographical variation or seasonal changes in food source could change the frequency of specific mitotypes. Is it possible then that dietary or thermal selection on random mitochondrial mutations that increase mitochondrial functions may lead to increased population subdivision and perhaps even speciation? Hill (2017) has championed the idea of an avian species concept that is based around mito-nuclear compatibility. The basic idea is that the mutation rate in mtDNA is generally higher than it is in nuclear-encoded genes, but the latter evolve to compensate for mtDNA mutations and restore OXPHOS function (Barreto and Burton 2013; Havird and Sloan 2016; Mishmar et al. 2006; Osada and Akashi 2012; van der Sluis et al. 2015). This species concept is an extension and expansion of cytonuclear species concept that has been presented in previous essays on the fundamental role of cytonuclear interactions in creating barriers to gene flow among populations (Bar-Yaacov et al. 2015; Burton and Barreto 2012; Chou and Leu 2010; Gershoni et al. 2009; Hill 2016; Lane 2009). Differences in dietary constraints support the theory that nutrition and food availability drive speciation (Yamagiwa and Basabose 2009). Many of the primate species in Southeast Asia rely heavily on mast fruiting events. Mast fruiting is the synchronous but variable production of large fruit crops (masts), which occurs throughout the world but differs in Malaysia because entire communities of unrelated species produce fruit during these events. This is called strict masting (Ashton et al. 1988; Visser et al. 2011), and it is important for numerous species in the ecosystems where this occurs including orangutans (*Pongo* spp.) (Brockman and van Schaik 2005) and two species of gibbon (*Hylobates muelleri* and *H. agilis*). These events are not predictable, although they seem to be tied to *El Nino* (Ashton et al. 1988; Wich and van Schaik 2000), but when they do occur, the species that exploit them gorge themselves and store the calories as fat (Knott 2005) because high-quality fall-back foods are not readily available in these geographic areas. From a mitochondrial perspective, the rate of the metabolism of these fat stores must occur at a rate that allows for survival (and reproduction) to the next mast event.

Further studies on the interactions between dietary macronutrients and specific mitotypes have implications in the fundamental disciplines of evolutionary biology and quantitative genetics where mtDNA is frequently used to monitor the movement of populations over space and time (Ballard and Kreitman 1995). Many of the hominid species in Garvin et al. (2015) that demonstrated positive selection at sites in the mtDNA-encoded ND5 subunit are either fruit eaters (frugivores) or leaf and bark eaters (folivores) confined to the tropics of either Southeast Asia or Africa (Brockman and van Schaik 2005). Dependence on fruits and leaves results in seasonal fluctuations of food. Large abundances of the primary food source are followed by scarcities, leaving animals to rely on lower-quality fall-back foods. Reliance on fall-back foods for prolonged periods results in reduced reproduction through delayed cycling and aborted fetuses, which could be a selective force that would drive mtDNA divergence in geographic regions with differing food availabilities (Altmann 2009). Consistent with this hypothesis, it has been argued that differences in fall-back foods drove population subdivision in Darwin's finches (Constantino and Wright 2009).

Studies of mitochondrial metabolic flexibility in model organisms and cell lines may also direct research in the medically applied fields of nutrigenomics and pharmacogenomics. MtDNA disorders were initially thought to be extremely rare in humans but are now known to affect ~1 in 4300 of the population (Gorman et al. 2015). Hundreds of different mtDNA mutations are associated with a wide range of overlapping clinical phenotypes. These disorders can be present from birth to old age and often affect the nervous system (Vafai and Mootha 2012). In parallel, somatic mutations of mtDNA have been shown to accumulate with age and to preferentially collect in organs affected in several common late-onset diseases (Wallace 2010). In many cases, the influence of a specific mutation can often only be detected when the proportion of the mutant mitotype reaches a threshold level that cannot be compensated by the cell (Stewart and Chinnery 2015). We suggest that future research may find novel and unexpected therapeutic strategies.

Acknowledgments We wish to thank Neil Youngson and Priscilla Gunadi for the comments. The review was supported by the Australian Research Council (ARC) Discovery Project 160102575.

References

Alberts B, Johnson A, Lewis J, Walter P, Raff M, Roberts K (2002) Molecular biology of the cell, 4th edn: international student edition. Routledge, San Francisco

Almeida LS, Nogueira C, Vilarinho L (2012) Nuclear-mitochondrial intergenomic communication disorders. In: Cseri J (ed) Skeletal muscle - from myogenesis to clinical relations. InTech, Croatia, pp 293–315

Altmann SA (2009) Fallback foods, eclectic omnivores, and the packaging problem. Am J Phys Anthropol 140:615–629

Andreyev AY, Kushnareva YE, Starkov AA (2005) Mitochondrial metabolism of reactive oxygen species. Biochemistry (Mosc) 70:200–214

Argyropoulos G, Harper ME (2002) Uncoupling proteins and thermoregulation. J Appl Physiol (1985) 92:2187–2198

Ashton PS, Givnish TJ, Appanah S (1988) Staggered flowering in the *Dipterocarpaceae*: new insights into the floral induction and the evolution of mast fruiting in the aseasonal tropics. Am Nat 132:44–66

Atkins PW (1994) The 2nd law: energy, chaos, and form. Henry Holt, New York

Aw WC, Correa CC, Clancy DJ, Ballard JWO (2011) Mitochondrial DNA variants in *Drosophila melanogaster* are expressed at the level of the organismal phenotype. Mitochondrion 11:756–763

Aw WC, Garvin MR, Melvin RG, Ballard JWO (2017) Sex-specific influences of mtDNA mitotype and diet on mitochondrial functions and physiological traits in *Drosophila melanogaster*. PLoS One 12:e0187554

Aw WC et al (2018) Genotype to phenotype: diet-by-mitochondrial DNA haplotype interactions drive metabolic flexibility and organismal fitness. PLoS Genet 14:e1007735

Bachman NJ, Yang TL, Dasen JS, Ernst RE, Lomax MI (1996) Phylogenetic footprinting of the human cytochrome c oxidase subunit VB promoter. Arch Biochem Biophys 333:152–162

Ballard JWO, James AC (2004) Differential fitness of mitochondrial DNA in perturbation cage studies correlates with global abundance and population history in *Drosophila simulans*. Proc R Soc Lond (Biol) 271:1197–1201

Ballard JWO, Kreitman M (1995) Is mitochondrial DNA a strictly neutral marker? Trends Ecol Evol 10:485–488

Ballard JWO, Pichaud N (2014) Mitochondrial DNA: more than an evolutionary bystander. Funct Ecol 28:218–231

Ballard JWO, Youngson NA (2015) Can diet influence the selective advantage of mitochondrial DNA haplotypes? Biosci Rep 52:15–19

Barreto FS, Burton RS (2013) Elevated oxidative damage is correlated with reduced fitness in interpopulation hybrids of a marine copepod. Proc Biol Sci 280:20131521

Bar-Yaacov D, Hadjivasiliou Z, Levin L, Barshad G, Zarivach R, Bouskila A, Mishmar D (2015) Mitochondrial involvement in vertebrate speciation? The case of mito-nuclear genetic divergence in chameleons. Genome Biol Evol 7:3322–3336

Bhopal RS, Rafnsson SB (2009) Could mitochondrial efficiency explain the susceptibility to adiposity, metabolic syndrome, diabetes and cardiovascular diseases in South Asian populations? Int J Epidemiol 38:1072–1081

Brand MD (2010) The sites and topology of mitochondrial superoxide production. Exp Gerontol 45:466–472

Brockman DK, van Schaik CP (2005) Seasonality in primates studies of living and extinct human and non-human primates. Cambridge University Press, New York

Burman JL et al (2014) A *Drosophila* model of mitochondrial disease caused by a complex I mutation that uncouples proton pumping from electron transfer. Dis Model Mech 7:1165–1174

Burton RS, Barreto FS (2012) A disproportionate role for mtDNA in Dobzhansky-Muller incompatibilities? Mol Ecol 21:4942–4957

Butow RA, Avadhani NG (2004) Mitochondrial signaling: the retrograde response. Mol Cell 14:1–15

Calabrese C et al (2013) Respiratory complex I is essential to induce a Warburg profile in mitochondria-defective tumor cells. Cancer Metab 1:1–15

Chae S et al (2013) A systems approach for decoding mitochondrial retrograde signaling pathways. Sci Signal 6:rs4

Chen S et al (2012) A cytoplasmic suppressor of a nuclear mutation affecting mitochondrial functions in *Drosophila*. Genetics 192:483–493

Chou JY, Leu JY (2010) Speciation through cytonuclear incompatibility: insights from yeast and implications for higher eukaryotes. BioEssays 32:401–411

Chou JY, Leu JY (2015) The Red Queen in mitochondria: cyto-nuclear co-evolution, hybrid breakdown and human disease. Front Genet 6:187

Clancy DJ (2008) Variation in mitochondrial genotype has substantial lifespan effects which may be modulated by nuclear background. Aging Cell 7:795–804
Cole MA et al (2011) A high fat diet increases mitochondrial fatty acid oxidation and uncoupling to decrease efficiency in rat heart. Basic Res Cardiol 106:447–457
Colombani J, Raisin S, Pantalacci S, Radimerski T, Montagne J, Leopold P (2003) A nutrient sensor mechanism controls *Drosophila* growth. Cell 114:739–749
Connallon T, Camus MF, Morrow EH, Dowling DK (2018) Coadaptation of mitochondrial and nuclear genes, and the cost of mother's curse. Proc Biol Sci 285: 20172257
Constantino PJ, Wright BW (2009) The importance of fallback foods in primate ecology and evolution. Am J Phys Anthropol 140:599–602
Correa CC, Aw WC, Melvin RG, Pichaud N, Ballard JWO (2012) Mitochondrial DNA variants influence mitochondrial bioenergetics in *Drosophila melanogaster*. Mitochondrion 12:459–464
Corton JM, Gillespie JG, Hardie DG (1994) Role of the AMP-activated protein kinase in the cellular stress response. Curr Biol 4:315–324
DeVivo DC, Leckie MP, Ferrendelli JS, McDougal DB Jr (1978) Chronic ketosis and cerebral metabolism. Ann Neurol 3:331–337
Ehinger JK et al (2016) Cell-permeable succinate prodrugs bypass mitochondrial complex I deficiency. Nat Commun 7:12317
Eleff S, Kennaway NG, Buist NR, Darley-Usmar VM, Capaldi RA, Bank WJ, Chance B (1984) 31P NMR study of improvement in oxidative phosphorylation by vitamins K3 and C in a patient with a defect in electron transport at complex III in skeletal muscle. Proc Natl Acad Sci U S A 81:3529–3533
Evans MJ, Scarpulla RC (1990) NRF-1: a trans-activator of nuclear-encoded respiratory genes in animal cells. Genes Dev 4:1023–1034
Fernández-Ayala DJM, Chen S, Kemppainen E, O'Dell KMC, Jacobs HT (2010) Gene expression in a *Drosophila* model of mitochondrial disease. PLoS One 5:e8549
Ferreira FL, Ladriere L, Vincent JL, Malaisse WJ (2000) Prolongation of survival time by infusion of succinic acid dimethyl ester in a caecal ligation and perforation model of sepsis. Horm Metab Res 32:335–336
Fujii N et al (2000) Exercise induces isoform-specific increase in 5'AMP-activated protein kinase activity in human skeletal muscle. Biochem Biophys Res Commun 273:1150–1155
Fumagalli M et al (2015) Greenlandic Inuit show genetic signatures of diet and climate adaptation. Science 349:1343–1347
Galluzzi L, Kroemer G (2014) Conceptual background and bioenergetic/mitochondrial aspects of oncometabolism, vol 542. Elsevier Science, Amsterdam
Garvin MR, Bielawski JP, Sazanov LA, Gharrett AJ (2015) Review and meta-analysis of natural selection in mitochondrial complex I in metazoans. J Zool Syst Evol Res 53:1–17
Geist V (1987) Bergmann's rule is invalid. Can J Zool 65:1035–1038
Gemmell NJ, Metcalf VJ, Allendorf FW (2004) Mother's curse: the effect of mtDNA on individual fitness and population viability. Trends Ecol Evol 19:238–244
Gershoni M, Templeton AR, Mishmar D (2009) Mitochondrial bioenergetics as a major motive force of speciation. BioEssays 31:642–650
Gomes AP et al (2013) Declining NAD(+) induces a pseudohypoxic state disrupting nuclear-mitochondrial communication during aging. Cell 155:1624–1638
Gorman GS et al (2015) Prevalence of nuclear and mitochondrial DNA mutations related to adult mitochondrial disease. Ann Neurol 77:753–759
Goto Y, Nonaka I, Horai S (1990) A mutation in the tRNA(Leu)(UUR) gene associated with the MELAS subgroup of mitochondrial encephalomyopathies. Nature 348:651–653
Guo R, Zong S, Wu M, Gu J, Yang M (2017) Architecture of human mitochondrial respiratory megacomplex I2III2IV2. Cell 170:1247–1257 e1212
Halliday R (1989) Food, reproduction and longevity: is the extended lifespan of calorie-restricted animals an evolutionary adaptation? BioEssays 10:125–127

Havird JC, Sloan DB (2016) The roles of mutation, selection, and expression in determining relative rates of evolution in mitochondrial versus nuclear genomes. Mol Biol Evol 33:3042–3053
Hill GE (2016) Mitonuclear coevolution as the genesis of speciation and the mitochondrial DNA barcode gap. Ecol Evol 6:5831–5842
Hill GE (2017) The mitonuclear compatibility species concept. Auk 134:393–409
Hirabara SM, Curi R, Maechler P (2010) Saturated fatty acid-induced insulin resistance is associated with mitochondrial dysfunction in skeletal muscle cells. J Cell Physiol 222:187–194
Hunt J, Brooks R, Jennions MD, Smith MJ, Bentsen CL, Bussiere LF (2004) High-quality male field crickets invest heavily in sexual display but die young. Nature 432:1024–1027
Ichiyama A (2011) Studies on a unique organelle localization of a liver enzyme, serine:pyruvate (or alanine:glyoxylate) aminotransferase. Proc Jpn Acad Ser B Phys Biol Sci 87:274–286
Innocenti P, Morrow EH, Dowling DK (2011) Experimental evidence supports a sex-specific selective sieve in mitochondrial genome evolution. Science 332:845–848
Jacinto E, Hall MN (2003) TOR signalling in bugs, brain and brawn. Nat Rev Mol Cell Biol 4:117–126
Jacob HS, Evans EW (2000) Influence of carbohydrate foods and mating on longevity of the parasitoid *Bathyplectes curculionis* (Hymenoptera: Ichneumonidae). Environ Entomol 29:1088–1095
Jacobs HT, Fernández-Ayala DJM, Manjiry S, Kemppainen E, Toivonen JM, O'Dell KMC (2004) Mitochondrial disease in flies. Biochim Biophys Acta 1659:190–196
James AC, Ballard JWO (2003) Mitochondrial genotype affects fitness in *Drosophila simulans*. Genetics 164:187–194
Jayasena T et al (2015) Upregulation of glycolytic enzymes, mitochondrial dysfunction and increased cytotoxicity in glial cells treated with Alzheimer's Disease plasma. PLoS One 10: e0116092
Jazwinski SM (2013) The retrograde response: when mitochondrial quality control is not enough. Biochim Biophys Acta 1833:400–409
Jheng HF et al (2012) Mitochondrial fission contributes to mitochondrial dysfunction and insulin resistance in skeletal muscle. Mol Cell Biol 32:309–319
Jornayvaz FR, Shulman GI (2010) Regulation of mitochondrial biogenesis. Essays Biochem 47:69–84
Kivisild T et al (2006) The role of selection in the evolution of human mitochondrial genomes. Genetics 172:373–387
Knott CD (2005) Energetic responses to food availability in the great apes: implications for hominin evolution. I. Seasonality in primates: studies of living and extinct human and non-human primates. Cambridge University Press, New York
Koppenol WH, Bounds PL, Dang CV (2011) Otto Warburg's contributions to current concepts of cancer metabolism. Nat Rev Cancer 11:325–337
Lane N (2009) Biodiversity: on the origin of bar codes. Nature 462:272–274
Lee HC, Wei YH (2000) Mitochondrial role in life and death of the cell. J Biomed Sci 7:2–15
Lee HC, Yin PH, Chi CW, Wei YH (2002) Increase in mitochondrial mass in human fibroblasts under oxidative stress and during replicative cell senescence. J Biomed Sci 9:517–526
Lee KP et al (2008) Lifespan and reproduction in *Drosophila*: new insights from nutritional geometry. Proc Natl Acad Sci U S A 105:2498–2503
Lee S-J, Zhang J, Choi AMK, Kim HP (2013) Mitochondrial dysfunction induces formation of lipid droplets as a generalized response to stress. Oxidative Med Cell Longev 2013:327167
Leigh SR (2004) Brain growth, life history, and cognition in primate and human evolution. Am J Primatol 62:139–164
Leonard WR, Robertson ML (1992) Nutritional requirements and human evolution: a bioenergetics model. Am J Hum Biol 4:179–195
Leonard WR, Robertson ML (1994) Evolutionary perspectives on human nutrition: the influence of brain and body size on diet and metabolism. Am J Hum Biol 6:77–88

Leonard WR, Sorensen MV, Galloway VA, Spencer GJ, Mosher MJ, Osipova L, Spitsyn VA (2002) Climatic influences on basal metabolic rates among circumpolar populations. Am J Hum Biol 14:609–620
Lionetti L et al (2014) High-lard and high-fish-oil diets differ in their effects on function and dynamic behaviour of rat hepatic mitochondria. PLoS One 9:e92753
Liu Y, Xu H, Yuan X, Rossiter SJ, Zhang S (2012) Multiple adaptive losses of alanine-glyoxylate aminotransferase mitochondrial targeting in fruit-eating bats. Mol Biol Evol 29:1507–1511
McCommis KS, Finck BN (2015) Mitochondrial pyruvate transport: a historical perspective and future research directions. Biochem J 466:443–454
McKenzie M, Lazarou M, Thorburn DR, Ryan MT (2007) Analysis of mitochondrial subunit assembly into respiratory chain complexes using Blue Native polyacrylamide gel electrophoresis. Anal Biochem 364:128–137
Michelakis ED (2008) Mitochondrial medicine: a new era in medicine opens new windows and brings new challenges. Circulation 117:2431–2434
Mihaylova MM, Shaw RJ (2011) The AMPK signalling pathway coordinates cell growth, autophagy and metabolism. Nat Cell Biol 13:1016–1023
Milot E, Moreau C, Gagnon A, Cohen AA, Brais B, Labuda D (2017) Mother's curse neutralizes natural selection against a human genetic disease over three centuries. Nat Ecol Evol 1:1400–1406
Mishmar D et al (2003) Natural selection shaped regional mtDNA variation in humans. Proc Natl Acad Sci U S A 100:171–176
Mishmar D, Ruiz-Pesini E, Mondragon-Palomino M, Procaccio V, Gaut B, Wallace DC (2006) Adaptive selection of mitochondrial complex I subunits during primate radiation. Gene 378:11–18
Morris G, Berk M (2015) The many roads to mitochondrial dysfunction in neuroimmune and neuropsychiatric disorders. BMC Med 13:1–24
Nelson DL, Lehninger AL, Cox MM (2008) Lehninger principles of biochemistry. W. H. Freeman, New York
Noguchi T (1987) Amino acid metabolism in animal peroxisomes. In: Fahimi HD, Sies H (eds) Peroxisomes in biology and medicine. Springer, Berlin, pp 234–243
O'Donnell J, Ahuja GD (2005) Drug injury: liability, analysis, and prevention. Lawyers & Judges, Tucson
Osada N, Akashi H (2012) Mitochondrial-nuclear interactions and accelerated compensatory evolution: evidence from the primate cytochrome C oxidase complex. Mol Biol Evol 29:337–346
Passos JF, von Zglinicki T, Kirkwood TB (2007) Mitochondria and ageing: winning and losing in the numbers game. BioEssays 29:908–917
Pichaud N, Messmer M, Correa CC, Ballard JWO (2013) Diet influences the intake target and mitochondrial functions of *Drosophila melanogaster* males. Mitochondrion 13:817–822
Ponce de Leon MS et al (2008) Neanderthal brain size at birth provides insights into the evolution of human life history. Proc Natl Acad Sci U S A 105:13764–13768
Protti A, Singer M (2006) Bench-to-bedside review: potential strategies to protect or reverse mitochondrial dysfunction in sepsis-induced organ failure. Crit Care 10:228
Quiros PM, Mottis A, Auwerx J (2016) Mitonuclear communication in homeostasis and stress. Nat Rev Mol Cell Biol 17:213–226
Rand DM, Haney RA, Fry AJ (2004) Cytonuclear coevolution: the genomics of cooperation. Trends Ecol Evol 19:645–653
Rho JM (2017) How does the ketogenic diet induce anti-seizure effects? Neurosci Lett 637:4–10
Ristow M, Schmeisser K (2014) Mitohormesis: promoting health and lifespan by increased levels of reactive oxygen species (ROS). Dose-Response 12:288–341
Robergs RA, Ghiasvand F, Parker D (2004) Biochemistry of exercise-induced metabolic acidosis. Am J Physiol Regul Integr Comp Physiol 287:R502–R516

Ruiz-Pesini E, Mishmar D, Brandon M, Procaccio V, Wallace DC (2004) Effects of purifying and adaptive selection on regional variation in human mtDNA. Science 303:223–226
Rutter J, Reick M, Wu LC, McKnight SL (2001) Regulation of clock and NPAS2 DNA binding by the redox state of NAD cofactors. Science 293:510–514
Sagan L (1967) On the origin of mitosing cells. J Theor Biol 14:255–274
Scarpulla RC (2008) Nuclear control of respiratory chain expression by nuclear respiratory factors and PGC-1-related coactivator. Ann N Y Acad Sci 1147:321–334
Schon EA, Manfredi G (2003) Neuronal degeneration and mitochondrial dysfunction. J Clin Invest 111:303–312
Schuelke M et al (1999) Mutant NDUFV1 subunit of mitochondrial complex I causes leukodystrophy and myoclonic epilepsy. Nat Genet 21:260–261
Selvi Rani D, Vanniarajan A, Gupta NJ, Chakravarty B, Singh L, Thangaraj K (2006) A novel missense mutation C11994T in the mitochondrial ND4 gene as a cause of low sperm motility in the Indian subcontinent. Fertil Steril 86:1783–1785
Semenza GL et al (2001) 'The metabolism of tumours': 70 years later. Novartis Found Symp 240:251–260 Discussion 260–254
Sena LA, Chandel NS (2012) Physiological roles of mitochondrial reactive oxygen species. Mol Cell 48:158–167
Senyilmaz D et al (2015) Regulation of mitochondrial morphology and function by stearoylation of TFR1. Nature 525:124–128
Sohal RS, Weindruch R (1996) Oxidative stress, caloric restriction, and aging. Science 273:59–63
Stewart JB, Chinnery PF (2015) The dynamics of mitochondrial DNA heteroplasmy: implications for human health and disease. Nat Rev Genet 16:530–542
Stringer CB, Andrews P (1988) Genetic and fossil evidence for the origin of modern humans. Science 239:1263–1268
Sugioka K, Nakano M, Totsune-Nakano H, Minakami H, Tero-Kubota S, Ikegami Y (1988) Mechanism of O2-generation in reduction and oxidation cycle of ubiquinones in a model of mitochondrial electron transport systems. Biochim Biophys Acta 936:377–385
Suliman HB, Carraway MS, Welty-Wolf KE, Whorton AR, Piantadosi CA (2003) Lipopolysaccharide stimulates mitochondrial biogenesis via activation of nuclear respiratory factor-1. J Biol Chem 278:41510–41518
Takayama T et al (2003) Control of oxalate formation from L-hydroxyproline in liver mitochondria. J Am Soc Nephrol 14:939–946
Tapia PC (2006) Sublethal mitochondrial stress with an attendant stoichiometric augmentation of reactive oxygen species may precipitate many of the beneficial alterations in cellular physiology produced by caloric restriction, intermittent fasting, exercise and dietary phytonutrients: "Mitohormesis" for health and vitality. Med Hypotheses 66:832–843
Thiriet M (2011) Cell and tissue organization in the circulatory and ventilatory systems. Springer, New York
Towarnicki SG, Ballard JWO (2017) *Drosophila* mitotypes determine developmental time in a diet and temperature dependent manner. J Insect Physiol 100:133–139
Trivers R (2002) Natural selection and social theory: selected papers of Robert Trivers. Oxford University Press, New York
Turrens JF, Alexandre A, Lehninger AL (1985) Ubisemiquinone is the electron donor for superoxide formation by complex III of heart mitochondria. Arch Biochem Biophys 237:408–414
Ulijaszek SJ (2002) Comparative energetics of primate fetal growth. Am J Hum Biol 14:603–608
Vafai SB, Mootha VK (2012) Mitochondrial disorders as windows into an ancient organelle. Nature 491:374–383
van de Weijer T et al (2013) Relationships between mitochondrial function and metabolic flexibility in Type 2 Diabetes mellitus. PLoS One 8:e51648
van der Sluis EO et al (2015) Parallel structural evolution of mitochondrial ribosomes and OXPHOS complexes. Genome Biol Evol 7:1235–1251

Visser MD, Jongejans E, van Breugel M, Zuidema PA, Chen Y-Y, Kassim AR, de Kroon H (2011) Strict mast fruiting for a tropical dipterocarp tree: a demographic cost–benefit analysis of delayed reproduction and seed predation. J Ecol 99:1033–1044
Wallace DC (2009) Mitochondria, bioenergetics, and the epigenome in eukaryotic and human evolution. Cold Spring Harb Symp Quant Biol 74:383–393
Wallace DC (2010) Colloquium paper: bioenergetics, the origins of complexity, and the ascent of man. Proc Natl Acad Sci U S A 107(Suppl 2):8947–8953
Warburg O, Posener K, Negelein E (1924) Uber den stoffwechsel der carcinomzelle. Biochem Z 152:309–344
Westermann B (2012) Bioenergetic role of mitochondrial fusion and fission. Biochim Biophys Acta 1817:1833–1838
Whelan SP, Zuckerbraun BS (2013) Mitochondrial signaling: forwards, backwards, and in between. Oxidative Med Cell Longev 2013:10
Wich SA, van Schaik CP (2000) The impact of El Nino on mast fruiting in Sumatra and elsewhere in Malesia. J Trop Ecol 16:563–577
Wojtaszewski JF, Nielsen P, Hansen BF, Richter EA, Kiens B (2000) Isoform-specific and exercise intensity-dependent activation of 5′-AMP-activated protein kinase in human skeletal muscle. J Physiol 528(Pt 1):221–226
Wu M, Gu J, Guo R, Huang Y, Yang M (2016) Structure of mammalian respiratory rupercomplex I1III2IV1. Cell 167:1598–1609 e1510
Yamagiwa J, Basabose AK (2009) Fallback foods and dietary partitioning among Pan and Gorilla. Am J Phys Anthropol 140:739–750
Ying W (2008) NAD+/NADH and NADP+/NADPH in cellular functions and cell death: regulation and biological consequences. Antioxid Redox Signal 10:179–206
Yoboue ED, Devin A (2012) Reactive oxygen species-mediated control of mitochondrial biogenesis. Int J Cell Biol 2012:403870
Youngson NA, Morris MJ, Ballard JWO (2017) The mechanisms mediating the antiepileptic effects of the ketogenic diet, and potential opportunities for improvement with metabolism-altering drugs. Seizure 52:15–19
Zera AJ, Harshman LG (2001) The physiology of life history trade-offs in animals. Annu Rev Ecol Syst 32:95–126
Zhang Y, Jiang L, Hu W, Zheng Q, Xiang W (2011) Mitochondrial dysfunction during in vitro hepatocyte steatosis is reversed by omega-3 fatty acid-induced up-regulation of mitofusin 2. Metabolism 60:767–775
Ziegler M (2000) New functions of a long-known molecule. Emerging roles of NAD in cellular signaling. Eur J Biochem 267:1550–1564

Transmission of Dysfunctional Mitochondrial DNA and Its Implications for Mammalian Reproduction

Kanokwan Srirattana and Justin C. St. John

Contents

Abstract Mitochondrial DNA (mtDNA) encodes proteins for the electron transport chain which produces the vast majority of cellular energy. MtDNA has its own replication and transcription machinery that relies on nuclear-encoded transcription and replication factors. MtDNA is inherited in a non-Mendelian fashion as maternal-only mtDNA is passed onto the next generation. Mutation to mtDNA can cause mitochondrial dysfunction, which affects energy production and tissue and organ function. In somatic cell nuclear transfer (SCNT), there is an issue with the mixing of two populations of mtDNA, namely from the donor cell and recipient oocyte. This review focuses on the transmission of mtDNA in SCNT embryos and offspring. The transmission of donor cell mtDNA can be prevented by depleting the donor cell of its mtDNA using mtDNA depletion agents prior to SCNT. As a result, SCNT embryos harbour oocyte-only mtDNA. Moreover, culturing SCNT embryos derived from

K. Srirattana and J. C. St. John (✉)
Mitochondrial Genetics Group, Hudson Institute of Medical Research, Clayton, VIC, Australia

Department of Molecular and Translational Sciences, Monash University, Clayton, VIC, Australia
e-mail: justin.stjohn@hudson.org.au

P. Sutovsky (ed.), *Cellular and Molecular Basis of Mitochondrial Inheritance*, Advances in Anatomy, Embryology and Cell Biology 231, https://doi.org/10.1007/102_2018_3, Published online: 9 January 2019

mtDNA depleted cells in media supplemented with a nuclear reprograming agent can increase the levels of expression of genes related to embryo development when compared with non-depleted cell-derived embryos. Furthermore, we have reviewed how mitochondrial supplementation in oocytes can have beneficial effects for SCNT embryos by increasing mtDNA copy number and the levels of expression of genes involved in energy production and decreasing the levels of expression of genes involved in embryonic cell death. Notably, there are beneficial effects of mtDNA supplementation over the use of nuclear reprograming agents in terms of regulating gene expression in embryos. Taken together, manipulating mtDNA in donor cells and/or oocytes prior to SCNT could enhance embryo production efficiency.

Keywords Embryo, Mitochondrial DNA, Mitochondrial supplementation, Replication, Somatic cell nuclear transfer, Transmission

1 Introduction

The mitochondrion is a double membrane bound organelle where energy is produced inside cells via a number of biochemical processes including the electron transport chain (ETC). Mitochondria have their own genome, mitochondrial DNA (mtDNA), which contains some of the protein-encoding genes of the ETC. However, mtDNA has a high mutation rate. Mutation to and deletion of the mitochondrial genome can lead to mitochondrial dysfunction which directly affects cellular energy production. As mtDNA is maternally inherited, these mutated mtDNA can pass through the oocyte to the next generation and cause mitochondrial disease. Assisted reproductive technologies such as somatic cell nuclear transfer (SCNT) have been widely used to produce embryos and offspring for agricultural and research purposes. However, there are several issues with SCNT, some of which are associated with mitochondria and mtDNA. In SCNT, a whole donor cell complete with its mtDNA is transferred into an enucleated oocyte. The mixing of two populations of mtDNA has been found in SCNT embryos, which can have negative effects on embryo development. We have reviewed how donor cell mtDNA can be transmitted to the embryo and how the transmission of donor cell mtDNA can be overcome. Moreover, we have also discussed new strategies to enhance embryo production by combining mitochondrial DNA supplementation with other assisted reproductive technologies. This would be beneficial not only for embryo production but also for a better understanding of nucleo-mitochondrial interactions.

2 Mitochondria and Mitochondrial DNA

Mitochondria are organelles that play important roles in energy production, cell death signalling, and the control of stress responses (reviewed by Vakifahmetoglu-Norberg et al. 2017). In cells, the production of adenosine triphosphate (ATP) is generated by

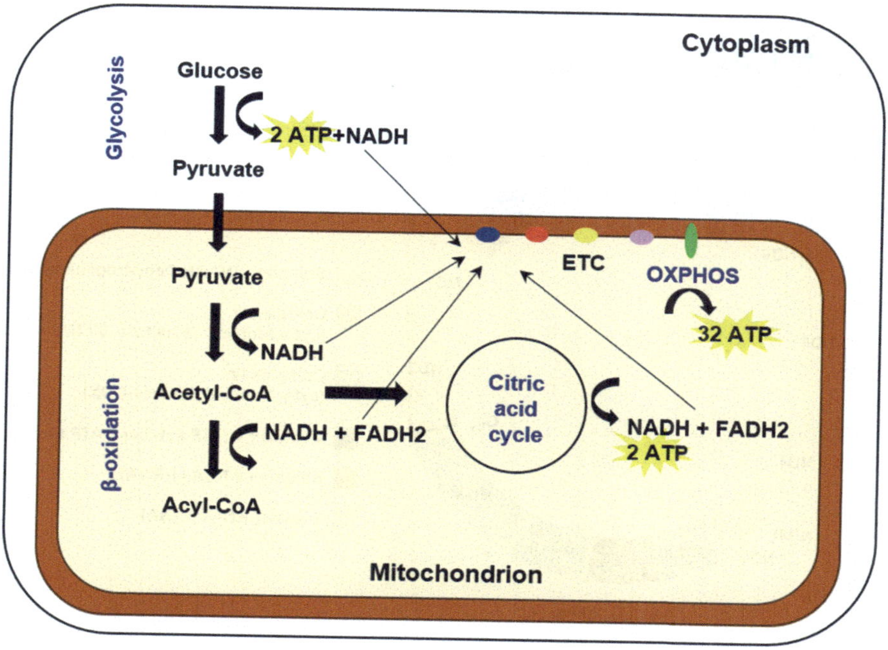

Fig. 1 Production of ATP. In the cytoplasm of the cell, glycolysis generates 2 molecules of ATP. In the mitochondrion, 2 molecules of ATP are generated by the citric acid cycle and 32 molecules are generated by OXPHOS (adapted from Pfeiffer et al. 2001)

glycolysis in the cell's cytoplasm, and β-oxidation, the citric acid cycle and oxidative phosphorylation (OXPHOS), which take place in the mitochondrion (Fig. 1). The majority of ATP in the cell is generated by OXPHOS (32 molecules of ATP) whilst glycolysis produces only two molecules of ATP (Pfeiffer et al. 2001). The mitochondrion has a double membrane, an outer membrane and an inner membrane, which houses the ETC (Capaldi 1982). The subunits of the ETC are encoded by both the mitochondrial and nuclear genomes (Attardi and Schatz 1988; Herrmann and Neupert 2000).

Mitochondrial DNA (mtDNA) is a circular, double-stranded molecule. It is 16.3 kb in size in cattle (Anderson et al. 1982), and approximately 16.6 kb in humans (Giles et al. 1980) and pigs (Ursing and Arnason 1998). The numbers of mitochondria and mtDNA copy number vary amongst cell types. For example, muscle cells have high numbers of mitochondria and possess between eight and ten copies of mtDNA per mitochondrion (reviewed by Moyes et al. 1998). On the other hand, primordial germ cells have about 200 mitochondria per cell and possess only one or two copies of mtDNA per mitochondrion (Cree et al. 2008). In each case, the numbers of mitochondria and copies of mtDNA per cell are indicative of a cell's requirement for ATP generated through OXPHOS where cells with higher numbers

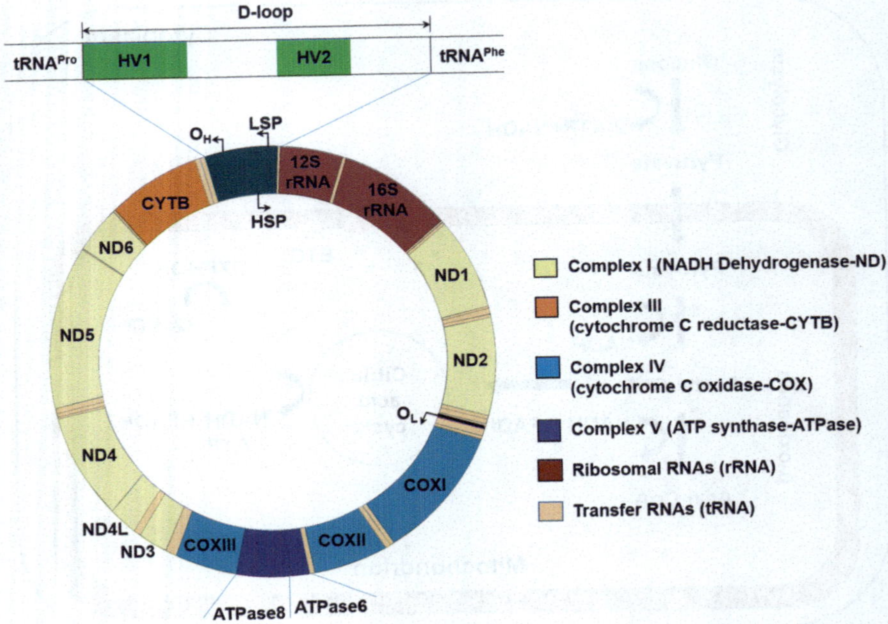

Fig. 2 The mammalian mitochondrial genome. MtDNA encodes 13 subunits of the electron transport chain. The genes associated with Complex I are ND 1, 2, 3, 4, 4 L, 5 and 6. Complex III consists of CYTB. Complex IV consists of COX I, II and III. The Complex V associated genes are ATPase 6 and 8. MtDNA also encodes the 12S and 16S rRNAs, and 22 tRNAs. The Displacement Loop (D-loop) is the control region where the heavy-strand promoter region (HSP), the light-strand promoter region (LSP), the origin of heavy-strand replication (O_H) and two hypervariable regions (HV1 and HV2) are located. Another control region is located between ND2 and COXI and is the site of the origin of light-strand replication (O_L) (adapted from Anderson et al. 1981)

are more likely to use OXPHOS and those with lower numbers utilise glycolysis (reviewed by Clay Montier et al. 2009; St. John et al. 2010).

MtDNA encodes 13 of the protein-encoding subunits of the ETC (Fig. 2), namely 7 subunits of NADH dehydrogenase (complex I), 1 subunit of cytochrome C reductase (complex III), 3 subunits of cytochrome C oxidase (complex IV) and 2 subunits of the ATP synthase (complex V) (Anderson et al. 1981). None of the subunits from complex II is encoded by mtDNA (Fig. 3). The majority of subunits are encoded by the nuclear genome. MtDNA also has 22 tRNAs, 2 rRNA genes and 1 non-coding region (the Displacement loop, D-loop) (Anderson et al. 1981). The D-loop is not only the control region, but it also possesses two hypervariable regions (HV1 and HV2) that are used by molecular geneticists to map the maternal ancestral inheritance of the mitochondrial genome (Torroni et al. 1994).

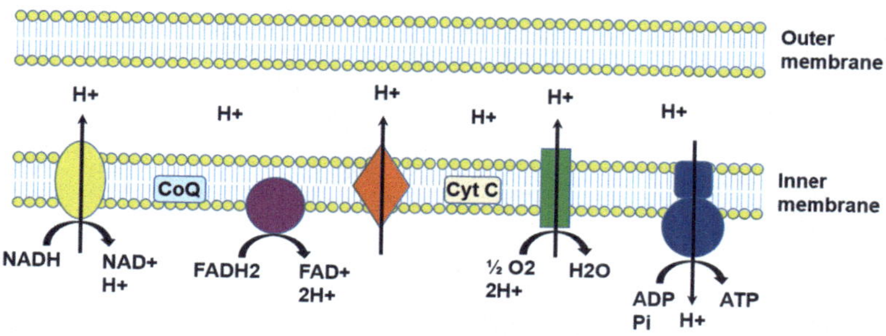

Complex	I	II	III	IV	V
Nuclear DNA	>18	4	8	10	10
mtDNA	7	0	1	3	2

Fig. 3 The electron transport chain. Electrons are derived from NADH and FADH2 and transferred to Complexes I (NADH dehydrogenase) and II (succinate dehydrogenase), respectively. Electrons are transferred to coenzyme Q (Co Q) and passed to Complex III (cytochrome c reductase). Cytochrome C (Cyt C) accepts electrons and donates them to Complex IV (cytochrome C oxidase). As electrons are transferred along Complex I, III and IV, energy is released and used to pump protons across the inner mitochondrial membrane to establish the mitochondrial membrane potential. The hydrogen ion gradient drives Complex V (ATPase) to generate ATP from adenosine diphosphate (ADP) and inorganic phosphate (Pi) (adapted from Pfeiffer et al. 2001)

3 Mitochondrial DNA Replication

Transcription and replication of mtDNA are initiated within the D-loop. MtDNA replication is controlled by nuclear-encoded transcription and replication factors that translocate to the mitochondrion and interact with the D-loop region of the mitochondrial genome to drive first transcription and then replication (Clayton 2000). MtDNA replication is initiated by mitochondrial transcription factor A (TFAM), which is supported by mitochondrial transcription factors B1 (TFB1M) and B2 (TFB2M) (Bonawitz et al. 2006). TFAM generates a small RNA-DNA hybrid fragment as a result of the cleavage of the long polycistronic transcript generated from transcription. This fragment acts as the primer for mtDNA replication (Chang and Clayton 1985). The mitochondrial-specific DNA polymerase, DNA polymerase gamma (POLG), then synthesises a new strand of mtDNA (Fig. 4) (Shadel and Clayton 1997). POLG consists of a catalytic subunit, DNA polymerase subunit gamma-1 (POLG1), which has 5′-3′ exonuclease activity, and an accessory subunit, DNA polymerase subunit gamma-2 (POLG2), which stabilises the catalytic enzyme (Gray and Wong 1992). The replication process is also supported by the mtDNA-specific helicase, Twinkle (TWNK), and the mitochondrial single-stranded DNA-binding protein (mtSSB) (Korhonen et al. 2003). mtDNA replication and

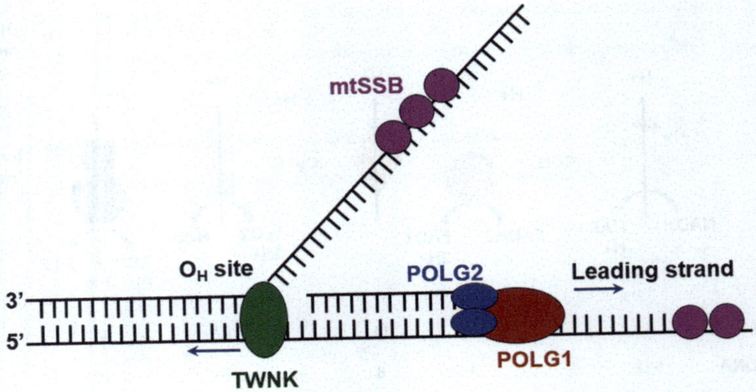

Fig. 4 Nuclear-encoded regulators of mtDNA replication. mtDNA replication is dependent on nuclear-encoded replication factors that translocate to the mitochondrion. The key mediators are mtSSB, TWNK, POLG1 and POLG2. POLG1 is the catalytic subunit that is DNA methylated, the levels of which correlate with mtDNA copy number and expression of POLG1 (adapted from Lee and St John 2015)

copy number are cell type specific (Kelly et al. 2012) with oocytes having the largest population of mtDNA of any cell type (Shoubridge 2000; Spikings et al. 2006).

Although the mode of mtDNA replication is still hotly debated, two models have been proposed. The first model is 'the strand-displacement model'. In this case, replication of mtDNA is initiated at the heavy (H) strand origin (O_H) in the D-loop of mtDNA (Clayton 1982). The synthesis of the H strand causes displacement of the H strand from the light (L) strand. Replication of the H strand progresses through two-thirds of the genome and exposes the L strand origin (O_L), and L strand synthesis is then initiated in the opposite direction. The replication of mtDNA in this manner does not involve the generation of Okazaki fragments, as for nuclear DNA. The second model is 'the strand-coupled model'. Shortly after replication of the H strand that begins at O_H, replication of the L strand is initiated from multiple sites and results in the generation of short Okazaki ribonucleotide fragments which are then converted to DNA (Holt et al. 2000). Previous reports have shown that both models exit and the occurrence may be dependent on cell type. The strand-displacement model is predominant in cells functioning at steady state levels. On the other hand, the strand-coupled model is predominate in cells that need to rapidly synthesise mtDNA such as the cells recovering from mtDNA depletion (Holt et al. 2000; Fish et al. 2004) and perhaps also during the early stages of development as cells first start to populate mtDNA to match their specific requirements for OXPHOS as they mature into adult cell types (Sun and St. John 2016).

4 mtDNA Replication During Development

In primordial germ cells, there are approximately 200 copies of mtDNA. mtDNA then increases in an exponential manner during oogenesis until it reaches a peak of approximately 200,000 copies at the metaphase II (MII) stage (Cao et al. 2007; Cree et al. 2008; Wai et al. 2008). In large mammalian species, such as the pig and cattle, mtDNA copy number gradually decreases during pre-implantation development (May-Panloup et al. 2005b; Spikings et al. 2007; Cagnone et al. 2016b). However, in the mouse, it remains fairly constant (Thundathil et al. 2005). At the blastocyst stage, mtDNA copy number increases only in the trophectodermal cells (Spikings et al. 2007), which are precursor cells that mediate implantation (reviewed by Brierley et al. 1998) and differentiate to form the placenta (Douglas et al. 2009). On the other hand, mtDNA copy number continues to progressively decrease in the inner cell mass cells (Spikings et al. 2007), which give rise to the embryo proper and the foetus, to establish the mtDNA set point (Facucho-Oliveira et al. 2007; Facucho-Oliveira and St. John 2009; Sun and St. John 2016). Indeed, after implantation, the inner cell mass cells differentiate into all cell types of the foetus, such as neurons, muscle, heart, blood and primordial germ cells, and mtDNA content varies amongst these cell types, depending on the specific requirements of those cells for ATP (Facucho-Oliveira et al. 2007; Kelly et al. 2012, 2013; St John 2012).

5 Mitochondrial DNA Transmission

The transmission of mtDNA takes place in a non-Mendelian manner with only mtDNA from the oocyte passed on to the next generation (Shoubridge and Wai 2007). Following normal fertilisation, the oocyte contributes to all of the mitochondria in the developing embryo as sperm mitochondria are normally eliminated through a process of ubiquitination (Sutovsky et al. 1999). The mixing of two populations of mtDNA, originating from the oocyte and sperm, can naturally occur in interspecies crossings, as shown in cattle (Sutovsky et al. 1999), mice (Gyllensten et al. 1991), sheep (Zhao et al. 2004), drosophila (Dokianakis and Ladoukakis 2014) and mussels (Zouros et al. 1992). However, sperm mitochondria are selectively destroyed during early embryo development in intra-specific crosses (Sutovsky et al. 1999).

Although mtDNA has a repair system, it is not sufficiently efficient at rescuing damage to mtDNA (Dianov et al. 2001; de Souza-Pinto et al. 2009). The mitochondrial genome has a high mutation rate, 10-fold to 20-fold-higher than for the nuclear genome (Shoffner and Wallace 1990). The patterns of mtDNA variation have been used to track human mutation (Ingman et al. 2000). A previous report using a mini-pig model has shown that non-pathogenic mtDNA variants can be transmitted from the mother to her daughter(s) (F1–F3); however, the levels of mtDNA variants differed across tissues, gender and generations (Cagnone et al. 2016a). Notably, non-pathogenic mutated mtDNA is often maintained at low levels in high ATP requiring tissues such as the heart, brain, diaphragm and muscle, which also have high mtDNA copy number and copy number increases to counter increased levels of these variants. Interestingly, mtDNA mutations also play important roles in mammalian aging (reviewed by Krishnan et al. 2007; Larsson 2010). In mouse, the rate of premature aging increases as the mtDNA mutation rate increases (Trifunovic et al. 2004; Kujoth et al. 2005), whilst a decrease in the mtDNA mutation rate extends the life span of an individual (Schriner et al. 2005). Whilst most naturally occurring mtDNA mutations are non-pathogenic (Ramos et al. 2013), specific nucleotide mutations and high levels of mutated mtDNA can cause mitochondrial disease (McFarland et al. 2007).

The first pathogenic mtDNA mutations were reported in 1988 (Holt et al. 1988; Wallace et al. 1988), and since then more than 250 pathogenic mtDNA mutations have been identified (see also MITOMAP: A Human Mitochondrial Genome Database, http://www.mitomap.org). The threshold of mutated mtDNA that leads to clinical symptoms varies for each mutation and amongst tissues. Normally, the pathogenic threshold of mutated mtDNA is around 60–90% mutated to wild-type mtDNA (Jeppesen et al. 2006; Wong 2007). The lower threshold applies to tissues that have a higher requirement for OXPHOS than tissues that primarily rely on glycolysis.

As mtDNA is maternally inherited, mutated mtDNA can be transmitted from the mother to her offspring. A previous report has shown that there is the possibility that both healthy and affected children can be produced from a healthy mother who harboured 50% mutated mtDNA (Larsson et al. 1992). The shift from a mixture of mutated and wild-type mtDNA to wild-type-only mtDNA between generations indicates the existence of an mtDNA genetic bottleneck (Upholt and Dawid 1977). The mechanisms regulating the mtDNA bottleneck are still debated. There are three possibilities for the occurrence of an mtDNA bottleneck. The first hypothesis suggests that the mtDNA bottleneck occurs during embryogenesis, which leads to a reduction in mtDNA copy number in the germ line (Jenuth et al. 1996; Cree et al. 2008). The second suggests that the mtDNA bottleneck is a result of the replication of a subgroup of mtDNA molecules during oogenesis, and not a reduction of mtDNA copy number in the germ line (Cao et al. 2007). The third hypothesis proposes that the mtDNA bottleneck results from the replication of a subpopulation of mtDNA during postnatal folliculogenesis (Wai et al. 2008).

Mutations to mtDNA can cause maternally inherited genetic diseases in early infancy or childhood such as Leigh syndrome (Finsterer 2008), depletion syndromes (Rahman and Poulton 2008), Kearns-Sayre syndrome (Maceluch and Niedziela 2006) and Pearson syndrome (Rötig et al. 1990). They can also occur in late childhood or adulthood and examples include Mitochondrial encephalopathy, lactic acidosis and stroke-like episodes (Goto et al. 1990), chronic progressive external ophthalmoplegia (Moraes et al. 1989), Leber's hereditary optic neuropathy (Man et al. 2002), and Myoclonic epilepsy and ragged red fibres (Shoffner et al. 1990; Yoneda et al. 1990). As there are no cures for mitochondrial disease, mitochondrial replacement therapies, such as pronuclear transfer, spindle transfer, cytoplasmic transfer and blastomere transfer, are, therefore, being applied to prevent the transmission of mutant mtDNA from the mother to her children (Tachibana et al. 2009; Zhang et al. 2017).

6 The Transmission of mtDNA in Somatic Cell Nuclear Transfer-Derived Embryos and Offspring

Somatic cell nuclear transfer (SCNT) involves the transfer of a nucleus from a donor cell into an enucleated oocyte of another individual (Wilmut et al. 1997). This technique provides a unique tool for the preservation of valuable individuals, propagation of livestock, genetically modified animal production and biomedical research (reviewed by Mastromonaco and King 2007), and can be used as a model for the study of the interactions between the nuclear and mitochondrial genomes (Srirattana and St. John 2018). There are still a number of confounding issues related to SCNT, especially the irregular patterns of transmission of the mitochondrial genome.

In SCNT, as a donor cell is transferred into an enucleated oocyte, not only is the nuclear genome passed onto the oocyte but also the mtDNA present in the mitochondria surrounding the donor cell can be transmitted (Sekirina et al. 1997; Takeda et al. 2003). Donor cells typically possess between 0.5 and 8×10^2 copies of mtDNA per cell (Burgstaller et al. 2007; Jiang et al. 2011) whilst cattle oocytes have between 3 and 6×10^5 copies of mtDNA (May-Panloup et al. 2005b; Iwata et al. 2011; Cree et al. 2015). There are three possibilities for the fate of donor cell mtDNA in SCNT-derived embryos: (1) mtDNA is only inherited from the oocyte whilst donor cell mtDNA is either not inherited or is eliminated; (2) mtDNA is inherited from the donor cell whilst oocyte mtDNA is eliminated; and (3) oocyte and donor cell mtDNA co-exist (Fig. 5). The mixing of donor cell and oocyte mtDNA could affect embryo development due to the incompatibility between the two genetically distinct sources of mtDNA co-existing that leads to inefficient mitochondrial respiration, dysfunctional electron transport chains and decreased levels of ATP (McKenzie and Trounce 2000; Steinborn et al. 2002; St. John et al. 2005).

From an initial report, the first SCNT-derived offspring from an adult mammary gland cell, Dolly, and also the other first set of nine SCNT-derived sheep inherited their mtDNA from the recipient oocyte and none of the donor cell's mtDNA was found in the tissues that were examined (Evans et al. 1999). Oocyte-only mtDNA was also found in cattle SCNT-derived embryos and offspring (Takeda et al. 1999). In 2000, the first report of the mixing of two populations of mtDNA in cattle SCNT-derived offspring was reported by Steinborn et al. The percentage of donor cell mtDNA in offspring ranged from 0.4 to 4% (Steinborn et al. 2000). The incidence of the persistence of donor cell mtDNA in SCNT-derived embryos and offspring has been listed in Table 1. Notably, high levels of donor cell mtDNA transmission

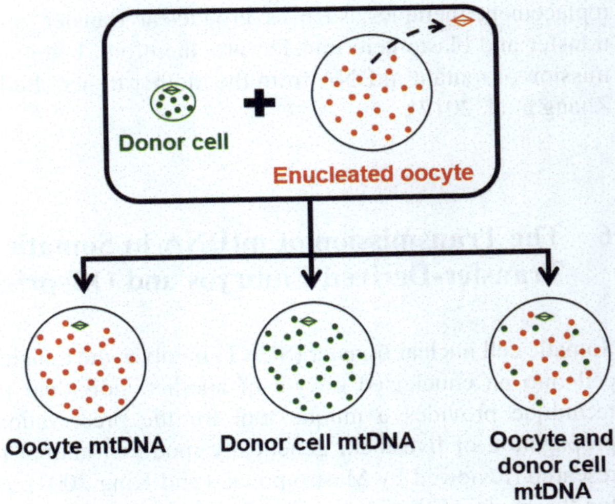

Fig. 5 Mitochondrial transmission in SCNT-derived embryos. There are three possible forms of mtDNA transmission in embryos: (1) the embryo harbours oocyte-only mtDNA (red enclosed circles); (2) the embryo harbours donor cell-only mtDNA (green enclosed circles); and (3) the embryo harbours oocyte and donor cell mtDNA (red and green enclosed circles)

Table 1 Donor cell mtDNA transmission following SCNT

Species	Stage of development	% donor cell mtDNA	Reference
Sheep	Offspring	0	Evans et al. (1999)
	Foetus and offspring	0.1–46.5	Burgstaller et al. (2007)
Cattle	Embryo and offspring	0	Takeda et al. (1999)
	Offspring	0.4–4.0	Steinborn et al. (2000)
	Embryos and offspring	0.6–12.7	Steinborn et al. (2002)
	Offspring	6.0–40.0	Takeda et al. (2003)
Mouse	Offspring	0–13.1	Inoue et al. (2004)
Porcine	Offspring	0.1–1.0	Takeda et al. (2006)
	Offspring	0–44.0	Do et al. (2012)

(40–47%) have been found in sheep, cattle and porcine offspring (Burgstaller et al. 2007; Takeda et al. 2003; Do et al. 2012). Moreover, in SCNT-derived cattle embryos produced by the handmade cloning method, where a donor cell is fused with one or more enucleated cytoplasts, 54 out of the 60 embryos, three out of the ten foetuses and three out of the six SCNT-derived offspring possessed donor cell mtDNA (Bowles et al. 2008).

Another model for studying mtDNA transmission in SCNT-derived embryos and offspring is interspecies SCNT where the donor cell from one species is transferred to an enucleated oocyte of another species. Lanza et al. (2000), who successfully produced interspecies SCNT gaur (*Bos gaurus*) using *Bos taurus* oocytes as recipients, reported that gaur interspecies SCNT-derived offspring harboured oocyte-only mtDNA (Lanza et al. 2000). Later Mastromonaco et al. found that there was both donor cell and oocyte mtDNA co-existing in interspecies SCNT gaur embryos (0.12–0.25% of donor cell mtDNA). Moreover, donor cell mtDNA copy number was constant during early preimplantation embryo development and increased at the blastocyst stage. However, this was not uniform as only four out of the 12 blastocysts showed an increase in donor cell mtDNA (Mastromonaco and King 2007). Similarly, the co-existence of donor cell and oocyte mtDNA was found in donor cell-cytoplast couplets (0.17% donor cell mtDNA) up to the blastocyst stage (0.03% donor cell mtDNA) in gaur interspecies SCNT embryos. However, it has been reported that donor cell mtDNA can persist in various tissues of gaur interspecies SCNT foetuses and offspring ranging from 0.03 to 99.0% (Imsoonthornruksa et al. 2012).

On the other hand, interspecies SCNT-derived offspring produced using coyote (*Canis latrans*) donor cells and domestic dog (*Canis familiaris*) oocytes harboured only oocyte mtDNA (Hwang et al. 2013). Chen et al. reported that mtDNA from both panda donor cells and rabbit oocytes co-existed at the blastocyst stage of panda-rabbit interspecies SCNT-derived embryos. Panda donor cell mtDNA was detected in foetuses, but the rabbit oocyte mtDNA had been eliminated (Chen et al. 2002). This suggests that the nuclear genome of the panda donor cell could support panda mitochondrial activity, but could not support the activity of the rabbit oocyte mitochondria. Notably, no live offspring were produced (Chen et al. 2002). The co-existence of donor cell and oocyte mtDNA was also found in goat-sheep interspecies SCNT (Bowles et al. 2007) and buffalo-cattle embryos (Srirattana et al. 2011). However, these interspecies SCNT-derived embryos could not progress beyond the 16-cell stage. Consequently, the success of viable interspecies SCNT-derived offspring was limited to genetically closely related species such as gaur-cattle (Lanza et al. 2000; Srirattana et al. 2012), mouflon-sheep (Loi et al. 2001), wild cat-domestic cat (Gómez et al. 2004), wolf-domestic dog (Kim et al. 2007), sand cat-domestic cat (Gómez et al. 2008) and ibex-goat (Folch et al. 2009) crosses.

A previous report has shown that the genetic distance between the donor cell and oocytes in each developmental stage from the 1-cell to offspring was 0.048–0.078% following bovine handmade cloning (Bowles et al. 2008). Following sheep SCNT, embryos that developed to the blastocyst stage had an mtDNA genetic divergence between the donor cell and recipient of oocyte of between 0.0022 and 0.0391%. On the other hand, the mtDNA divergence of arrested embryos ranged between 0.0112 and 0.0272% (Bowles et al. 2007). Thus, it appears that the genetic distance between the donor cell and recipient oocyte is the key to the success of embryo production. A little increase of genetic distance between the donor cell and oocyte can increase the developmental potential. However, a more pronounced increase in genetic distance, as with interspecies cloning, reduces success rates for SCNT (reviewed by Kelly and St. John 2011).

Furthermore, in mouse models, it is well documented that two genetically distinct populations of mtDNA can lead to severe metabolic syndromes and other disorders (Nakada et al. 2001; Acton et al. 2007). Many of these disorders are similar to those reported in SCNT-derived animals (Cibelli et al. 2002). In addition, it is now well documented that the mixing of two populations of mtDNA likely affects embryo development (Takeda et al. 2005; Lloyd et al. 2006; Bowles et al. 2007; Chen et al. 2002; Thongphakdee et al. 2008; Jiang et al. 2011) probably resulting from the nuclear-encoded proteins of the electron transport chain interacting with two distinct populations of proteins encoded by the two genetically distinct populations of mtDNA (McKenzie and Trounce 2000; Steinborn et al. 2002; St. John et al. 2005). However, it is not only the mixing of donor cell and oocyte mtDNA that could lead to developmental abnormalities, the donor cell could be contributing aged and damaged mitochondria to the oocyte, which could trigger processes such as apoptosis in the developing embryo (Li et al. 2000; Zhao et al. 2015) and lead to the occurrence of late-onset disease. Likewise, mitochondria isolated from somatic cells are metabolically more active (Takeda et al. 2010), which could be detrimental to early developmental outcomes. Somatic cell mitochondria often have the potential to initiate key cellular signalling processes including apoptosis (reviewed by Tait and Green 2010). Furthermore, the mtDNA present in somatic cells is likely to have accumulated deletions or mutations as part of the aging process (Cortopassi et al. 1992; Brierley et al. 1998), which, if preferentially replicated during development, could adversely affect the health, well-being and performance of the resultant offspring. Indeed, it has been demonstrated that somatic cell mitochondria can have negative effects on embryo development (Takeda et al. 2005, 2010).

7 How Is Donor Cell mtDNA Transmitted?

As oocytes mature to the metaphase II stage, replication of mtDNA ceases and mtDNA replication does not take place until later in development (please refer above) (Lloyd et al. 2006; Bowles et al. 2007). However, somatic donor cells continue to express genes involved in mtDNA replication such as POLG and TFAM prior to SCNT (Lloyd et al. 2006). Indeed, they continue to be expressed in SCNT-derived preimplantation embryos until they reach the blastocyst stage (Bowles et al. 2007). These results suggest a reason as to why donor cell mtDNA persists from donor cell transfer into preimplantation development. Moreover, the expression levels of the mtDNA replication factors in SCNT-derived embryos were higher than that of IVF-derived embryos (Lloyd et al. 2006; Bowles et al. 2007). Notably, the transmission of donor cell mtDNA appears to be random and is not dependent on the amount of donor cell mtDNA that is introduced into an enucleated oocyte (Lloyd et al. 2006).

8 Overcoming the Transmission of Donor Cell mtDNA

The issue with the mixing of two populations of mtDNA in SCNT-derived embryos and offspring can be overcome by depleting the donor cell of its mtDNA prior to SCNT by using, for example, mtDNA depletion agents, such as ethidium bromide (Desjardins et al. 1985; Hayakawa et al. 1998; Lloyd et al. 2006; Lee et al. 2010a) or 2′,3′-dideoxycytidine (ddC) (Brinkman and Kakuda 2000). In these previous reports, 50 ng/mL ethidium bromide was used to deplete mtDNA from donor cells (Lloyd et al. 2006; Lee et al. 2010b). However, ethidium bromide is a potentially highly

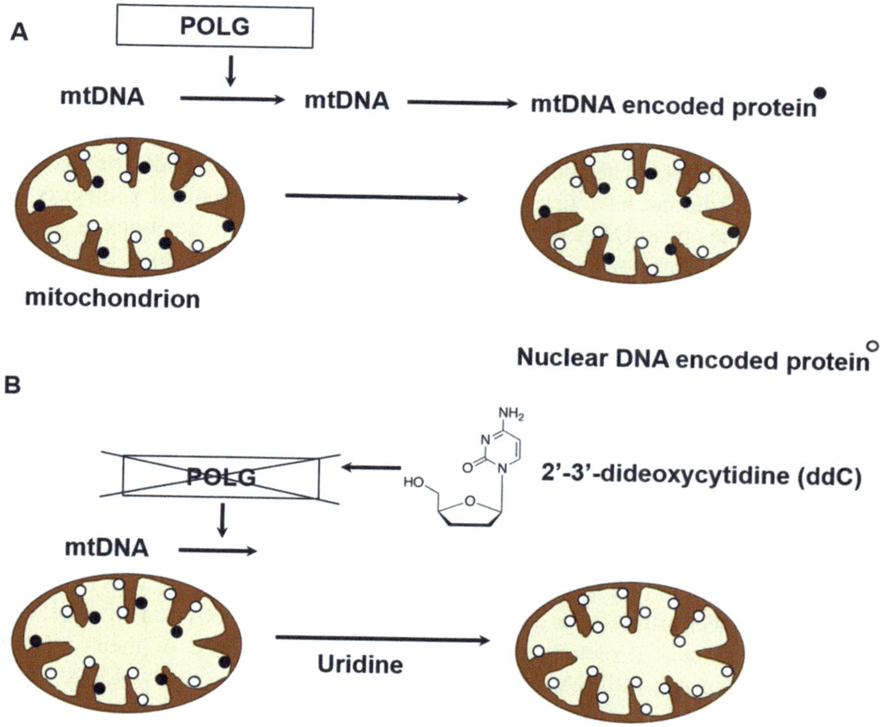

Fig. 6 The effects of ddC on mtDNA replication. (**a**) Normal replication of mtDNA is regulated by POLG. (**b**) ddC inhibits POLG from interacting with the mitochondrial genome. Consequently, as mtDNA replication does not take place, the cell's mtDNA content is diluted out with each cell division whilst in culture. Uridine is added as a substrate to promote pyrimidine synthesis (adapted from Brinkman and Kakuda 2000)

toxic mutagen and also a possible carcinogen or teratogen (Waring 1965) but less likely at low concentrations. The nucleoside reverse transcriptase inhibitor, ddC, has been widely used to suppress HIV replication and affects the interaction of POLG with the mitochondrial genome (Brinkman and Kakuda 2000). Normal replication of mtDNA is induced by POLG, resulting in sufficient copies of mtDNA for transcription and mtDNA-encoded protein production through translation and, thus, functional mitochondria. When ddC inhibits POLG, mtDNA replication and mtDNA-encoded protein production are stopped. Consequently, as mtDNA replication does not take place, the cell's mtDNA content is diluted out with each cell division whilst in culture (Brinkman et al. 1999) (Fig. 6). Moreover, the actions of ddC are more specific as an inhibitor of mtDNA replication than its counterpart ethidium bromide (Brown and Clayton 2002). Indeed, ddC is specific to POLG as it does not affect the polymerases that drive replication of nuclear DNA (e.g. DNA polymerase alpha, delta and epsilon; reviewed by Brinkman et al. 1998). The loss of functional mitochondrial electron transport chains in mtDNA depleted cells inhibits pyrimidine synthesis resulting in the failure of cells to proliferate (King and Attardi 1989). To resolve this problem, uridine is added to the culture medium as the substrate for pyrimidine synthesis (Grégoire et al. 1984), which has been widely used in ethidium bromide (Lloyd et al. 2006; Lee et al. 2010b) and ddC experiments (Keilbaugh et al. 1993; Nelson et al. 1997). From our previous report, the donor cells were not able to replenish their mtDNA when cultured in media supplemented with ddC for 30 days. Moreover, ddC did not have detrimental effects on chromosomal DNA integrity as the chromosomes of mtDNA depleted cells did not exhibit any deletions or duplications when compared with non-depleted cells (Srirattana and St. John 2017).

9 Enhancing SCNT Efficiency by Manipulating Donor Cell mtDNA

As mentioned earlier, the mixing of donor cell and oocyte mtDNA had negative effects on embryo developmental potential. The most appropriate way to ensure that donor cell mtDNA is not passed onto the embryo is to deplete the donor cell of its mtDNA prior to SCNT by using an mtDNA depletion agent. In sheep, SCNT-derived embryos produced from donor cells possessing their full mtDNA complement, and donor cells partially depleted and almost completely depleted of their mtDNA harboured donor cell mtDNA to 0.90, 0.57 and 0.01%, respectively (Lloyd et al. 2006), resulting in the continued persistence of donor cell mtDNA in embryos. Later, Lee et al. successfully produced SCNT-derived offspring that harboured oocyte-only mtDNA by SCNT using mtDNA depleted cells as the donor cells. In contrast, the offspring produced by non-depleted cells harboured both donor cell and oocyte mtDNA (Lee et al. 2010a). Depleting mtDNA from the donor cells prior to SCNT ensures that the resultant embryo and offspring transmit only oocyte mtDNA and, thus, this mimics the patterns of normal mtDNA inheritance consistent with the less invasive assisted reproductive technologies, such as in vitro fertilisation (Sutovsky et al. 1999), and natural conception (Giles et al. 1980). In 2017, our group used ddC to deplete mtDNA from cattle cells and the mtDNA depleted cells were used as the donor cells for SCNT embryo production (Fig. 7) (Srirattana and St. John 2017). All SCNT-derived embryos produced from mtDNA depleted cells harboured oocyte-only mtDNA. On the other hand, the mixing of donor cell and recipient oocyte mtDNA was found in embryos produced from non-depleted cells (0–30% donor cell mtDNA). The persistence of donor cell mtDNA has been reported in cattle SCNT-derived embryos (Meirelles et al. 2001) and offspring (Steinborn et al. 2002; Hiendleder et al. 2003; Takeda et al. 2003) when non-depleted cells were used as the donor cells. Moreover, depleting mtDNA from donor cells did not have detrimental effects on the quality of embryos in terms of cell number and mtDNA copy number at the blastocyst stage, and also increased levels of expression of genes involved in DNA repair. Notably, the reprogramming agent, trichostatin A (TSA), can enhance blastocyst rates for embryos produced from depleted cells to levels equivalent to that

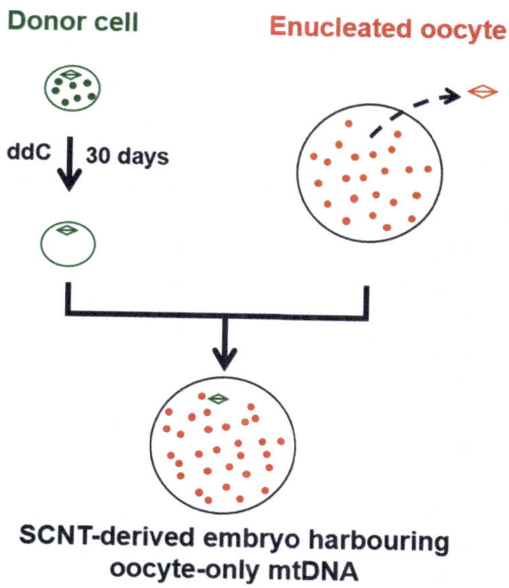

Fig. 7 SCNT-derived embryo production. MtDNA is depleted from the donor cells (green enclosed circles) by culturing in culture media supplemented with ddC for 30 days. Then, an mtDNA depleted cell is transferred into an enucleated oocyte. As a result, the embryo harbours oocyte-only mtDNA (red enclosed circles)

of non-depleted cells. The combination of depleting mtDNA from donor cells and the use of TSA positively modulated the expression levels of genes involved in DNA methylation, and embryonic formation and development. Taken together, depleting mtDNA from donor cells prior to SCNT not only prevents the transmission of donor cell mtDNA but also positively modulates the gene expression patterns of SCNT-derived embryos (Srirattana and St. John 2017).

10 The Effects of mtDNA Copy Number on Oocytes and Embryo Development

MtDNA copy number in oocytes affects fertilisation and embryo development in mice (Thundathil et al. 2005), pigs (El Shourbagy et al. 2006; Spikings et al. 2007; Cagnone et al. 2016b), cattle (May-Panloup et al. 2005b; Takeo et al. 2013) and humans (Reynier et al. 2001; May-Panloup et al. 2005a; Santos et al. 2006). Oocytes with low mtDNA copy number and insufficient levels of ATP are less likely to develop to the blastocyst stage (van Blerkom et al. 1995; Wakefield et al. 2011; Cagnone et al. 2016b). Moreover, mtDNA copy number in oocytes also affects post-implantation development (Wai et al. 2010). Replication of mtDNA in embryos does not take place until the blastocyst stage and is limited only to the trophectodermal cells from which the placenta is derived, and does not take place in the inner cell mass (May-Panloup et al. 2005a; Spikings et al. 2007). Therefore, sufficient mtDNA in the oocyte at the time of fertilisation is important to support embryo development to term.

Staining with the dye Brilliant Cresyl Blue (BCB) was first used as a marker to pre-select porcine oocytes based on their developmental potential (Ericsson et al. 1993). Growing oocytes have high levels of glucose-6-phosphate dehydrogenase (G6PDH) activity, which can reduce BCB to a colourless state (BCB^-). On the other hand, G6PDH activity is inactive in oocytes at the end of their growth phase. G6PDH in these oocytes cannot reduce BCB and these oocytes present with blue cytoplasms and are, therefore, designated BCB^+ (Mangia and Epstein 1975). BCB staining has been widely used for selecting good quality oocytes in many species such as pig (Ericsson et al. 1993; El Shourbagy et al. 2006; Fu et al. 2015; Cagnone et al. 2016b), cattle (Pujol et al. 2004; Bhojwani et al. 2007; Sugulle et al. 2008; Su et al. 2012; Castaneda et al. 2013; Fakruzzaman et al. 2013; Silva et al. 2013; Salviano et al. 2015), horses (Mohammadi-Sangcheshmeh et al. 2011), mice (Wu et al. 2007), buffalos (Manjunatha et al. 2007), dogs (Rodrigues et al. 2009), rats (Alcoba et al. 2011), sheep (Mohammadi-Sangcheshmeh et al. 2012; Wang et al. 2012), camels (Fathi et al. 2017) and humans (Duarte Alcoba et al. 2017). BCB^+ oocytes have higher maturation (Mohammadi-Sangcheshmeh et al. 2011; Fu et al. 2015), fertilisation (El Shourbagy et al. 2006), blastocyst (Salviano et al. 2015; Cagnone et al. 2016b) and calving rates (Su et al. 2012) when compared with BCB^- oocytes. Moreover, BCB^+ oocytes have higher mtDNA copy number when compared with BCB^- oocytes (El Shourbagy et al. 2006; Cagnone et al. 2016b).

Mitochondrial supplementation has been shown to positively increase blastocyst rates and the quality of embryos in mice (Pinkert et al. 1997; van Blerkom et al. 1998; Takeda et al. 2005, 2010; Yi et al. 2007), humans (Tzeng et al. 2004), cattle (Hua et al. 2007; Ferreira et al. 2010) and pigs (Cagnone et al. 2016b; El Shourbagy et al. 2006) when their oocytes had low mtDNA copy number. However, it is important to study how additional mtDNA interacts with the existing nuclear and mitochondrial genomes. SCNT can be used as a model to study nuclear and mitochondrial genome interactions as this technique can produce genetically identical embryos. Recently, we produced cattle embryos using mitochondrial supplementation combined with SCNT, otherwise named as miNT (Srirattana and St. John 2018). BCB^+ oocytes were used as the recipient oocytes to ensure that the oocyte had sufficient copies of mtDNA, and mitochondria extracted from BCB^+ oocytes (Cagnone et al. 2016b), approximately 560 copies of mtDNA, were injected into the enucleated oocyte. mtDNA depleted cells were used as the donor cells to eliminate the detrimental effects of donor cell mtDNA (Fig. 8).

In miNT-derived embryos, depleting mtDNA from donor cells increased the levels of expression of genes involved in embryonic development. Additional mtDNA did not increase blastocyst rates in miNT-derived embryos, similar to a previous report in the pig when oocytes with sufficient mtDNA (BCB^+ oocytes) were supplemented with mtDNA (Cagnone et al. 2016b). Although cell numbers at the blastocyst stage of miNT-derived embryos were slightly lower, their mtDNA copy number was two times higher than that for SCNT-derived embryos. Moreover, gene expression patterns of miNT-derived embryos were very different when compared with

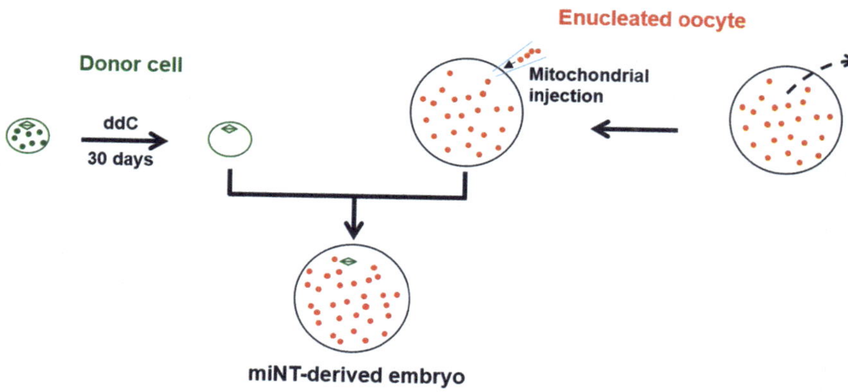

Fig. 8 miNT-derived embryo production. Mitochondria are extracted from oocytes (red enclosed circles) and then injected into an enucleated oocyte. A mtDNA-depleted cell is then transferred into the recipient oocyte. The resultant embryo harbours oocyte-only mtDNA (red enclosed circles)

SCNT-derived embryos. Additional mtDNA increased expression levels of genes involved in oxidative phosphorylation, cell cycle and DNA repair; however, it decreased expression levels of genes involved in cell survival, DNA replication and post-translational modification. Notably, the use of TSA did not enhance embryo development or change gene expression patterns of miNT-derived embryos, unlike SCNT-derived embryos (Srirattana and St. John 2018).

Moreover, miNT-derived embryos produced from depleted cells cultured in the absence or presence of TSA were compared with SCNT-derived embryos produced from depleted cells in the presence of TSA, which were assumed to be a gold standard, as the use of mtDNA depleted donor cells and TSA increased developmental rates and expression levels of genes involved in embryonic development in SCNT-derived embryos (Srirattana and St. John 2017). Additional mtDNA alone increased expression levels of genes involved in glycolysis and decreased levels of expression of genes involved in embryonic death in miNT-derived embryos produced from depleted cells in the absence of TSA when compared with SCNT-derived embryos produced from depleted cells in the presence of TSA. Notably, the combination of additional mtDNA and the use of TSA had negative effects in terms of increasing expression levels of genes involved in embryonic death and decreased expression levels of genes involved in cell cycle and cell morphology in miNT-derived embryos. Taken together, this study showed that additional mtDNA had beneficial effects over the use of TSA suggesting that it can mediate early reprogramming effects. This is the first report showing how additional mtDNA influences embryo development and gene expression profiles of SCNT embryos. The effects of manipulating mtDNA on embryos have been summarised in Table 2.

Table 2 The effects of manipulating mtDNA on preimplantation embryo development

Effects of	Blastocyst rate	mtDNA copy number per cell in blastocysts	Effects on gene expression levels
Donor cell mtDNA depletion	↓[a]	=	↑ DNA repair
Donor cell mtDNA depletion + TSA	↑[a]	=	↑ Embryonic development
Additional mtDNA	=	↑[a]	↑ Energy production ↓ Embryonic cell death
Additional mtDNA + TSA	=	↑[a]	↑ Cellular function and maintenance ↑ Embryonic death

[a]Significant difference

11 mtDNA and Other Assisted Reproductive Technologies

Assisted reproductive technologies such as in vitro fertilisation (IVF), intracytoplasmic sperm injection (ICSI) and cytoplasmic transfer (CT) have been used to resolve male and female infertility. ICSI is a technique that involves the injection of a single sperm into the recipient oocyte to overcome male infertility (Palermo et al. 1992). Although a whole sperm including its head, midpiece and tail are introduced into the oocyte, there appears to be no transmission of sperm mtDNA to the offspring (Houshmand et al. 1997; Danan et al. 1999; Marchington et al. 2002).

Another technique is CT, also known as ooplasmic transfer, which involves injecting a single sperm and about 5–15% cytoplasm from a donor oocyte into the recipient oocyte (Brenner et al. 1997; Cohen et al. 1997). This technique has been applied to patients who have poor embryonic development and experienced implantation failure after several rounds of IVF-derived embryo transfer. The first report of offspring derived using this technique was in 1997 (Cohen et al. 1997). However, the children born through CT harboured mtDNA from both the donor and recipient oocytes (Brenner et al. 2000). The co-existence of donor and recipient oocyte mtDNA has been reported in other species such as mouse (Sharpley et al. 2012) and cattle (Ferreira et al. 2010). Moreover, a previous report has shown that the donor oocyte alters the epigenetic processing of the paternal genome (Liang et al. 2009). The mixing of two populations of mtDNA from donor and recipient oocytes in mouse models has been shown to affect the physiological functions of offspring (Acton et al. 2007). Another technique, blastomere transfer, has been developed as a method to prevent mtDNA disorders, which involves fusing a blastomere from an embryo harbouring mutant mtDNA into an enucleated donor oocyte harbouring wild-type mtDNA. The cattle embryos produced by this technique harboured both mutant and wild-type mtDNA (Steinborn et al. 2000; Ferreira et al. 2007). The outcome is similar to cytoplasmic transfer where the embryos inherit two populations of mtDNA. However, this technique has not been applied clinically yet.

12 Conclusions

In this review, we have shown that the transmission of dysfunctional mtDNA could be prevented by depleting mtDNA from the donor cell prior to SCNT. Mitochondrial supplementation in oocytes not only enhances the efficiency of embryo production by increasing mtDNA copy number and the expression levels of genes in involved in energy production and DNA repair, but it also provides insights into the importance of the interactions between the nuclear and mitochondrial genomes. The combination of improving knowledge about mtDNA and assisted reproductive technologies such as manipulating donor cell mtDNA and/or supplementing oocytes with mitochondria will result in better quality and numbers of embryo produced.

Acknowledgements K.S. was supported by a Monash Graduate Scholarship and a Monash International Postgraduate Research Scholarship.

References

Acton BM, Lai I, Shang X, Jurisicova A, Casper RF (2007) Neutral mitochondrial heteroplasmy alters physiological function in mice. Biol Reprod 77(3):569–576. https://doi.org/10.1095/biolreprod.107.060806

Alcoba DD, da Rosa Braga BL, Sandi-Monroy NL, Proença LA, Felix Lopes RF, de Oliveira ATD (2011) Selection of Rattus norvegicus oocytes for in vitro maturation by brilliant cresyl blue staining. Zygote 21(3):238–245. https://doi.org/10.1017/S0967199411000463

Anderson S, Bankier A, Barrell B, de Bruijn M, Coulson A, Drouin J, Eperon I, Nierlich D, Roe B, Sanger F, Schreier P, Smith A, Staden R, Young I (1981) Sequence and organization of the human mitochondrial genome. Nature 290(5806):457–465. https://doi.org/10.1038/290457a0

Anderson S, de Bruijn MHL, Coulson AR, Eperon IC, Sanger F, Young IG (1982) Complete sequence of bovine mitochondrial DNA conserved features of the mammalian mitochondrial genome. J Mol Biol 156(4):683–717. https://doi.org/10.1016/0022-2836(82)90137-1

Attardi G, Schatz G (1988) Biogenesis of mitochondria. Annu Rev Cell Biol 4:289–333

Bhojwani S, Alm H, Torner H, Kanitz W, Poehland R (2007) Selection of developmentally competent oocytes through brilliant cresyl blue stain enhances blastocyst development rate after bovine nuclear transfer. Theriogenology 67(2):341–345. https://doi.org/10.1016/j.theriogenology.2006.08.006

Bonawitz N, Clayton D, Shadel G (2006) Initiation and beyond: multiple functions of the human mitochondrial transcription machinery. Mol Cell 24(6):813–825. https://doi.org/10.1016/j.molcel.2006.11.024

Bowles EJ, Lee JH, Alberio R, Lloyd REI, Stekel D, Campbell KHS, St. John JC (2007) Contrasting effects of in vitro fertilization and nuclear transfer on the expression of mtDNA replication factors. Genetics 176(3):1511–1526. https://doi.org/10.1534/genetics.106.070177

Bowles EJ, Tecirlioglu RT, French AJ, Holland MK, St. John JC (2008) Mitochondrial DNA transmission and transcription after somatic cell fusion to one or more cytoplasts. Stem Cells 26 (3):775–782. https://doi.org/10.1634/stemcells.2007-0747

Brenner CA, Scott RT Jr, Cohen J (1997) Use of enucleate donor oocyte cytoplasm in recipient eggs. Lancet 350(9082):961–962. https://doi.org/10.1016/S0140-6736(05)63306-5

Brenner CA, Barritt JA, Willadsen S, Cohen J (2000) Mitochondrial DNA heteroplasmy after human ooplasmic transplantation. Fertil Steril 74(3):573–578. https://doi.org/10.1016/S0015-0282(00)00681-6

Brierley EJ, Johnson MA, Lightowlers RN, James OFW, Turnbull DM (1998) Role of mitochondrial DNA mutations in human aging: implications for the central nervous system and muscle. Ann Neurol 43(2):217–223. https://doi.org/10.1002/ana.410430212

Brinkman K, Kakuda T (2000) Mitochondrial toxicity of nucleoside analogue reverse transcriptase inhibitors: a looming obstacle for long-term antiretroviral therapy? Curr Opin Infect Dis 13 (1):5–11

Brinkman K, ter Hofstede H, Burger D, Smeitink J, Koopmans P (1998) Adverse effects of reverse transcriptase inhibitors: mitochondrial toxicity as common pathway. AIDS (London, England) 12(14):1735–1744

Brinkman K, Smeitink J, Romijn J, Reiss P (1999) Mitochondrial toxicity induced by nucleoside-analogue reverse-transcriptase inhibitors is a key factor in the pathogenesis of antiretroviral-therapy-related lipodystrophy. Lancet 354(9184):1112–1115. https://doi.org/10.1016/S0140-6736(99)06102-4

Brown TA, Clayton DA (2002) Release of replication termination controls mitochondrial DNA copy number after depletion with 2′,3′-dideoxycytidine. Nucleic Acids Res 30(9):2004–2010

Burgstaller J, Schinogl P, Dinnyes A, Müller M, Steinborn R (2007) Mitochondrial DNA heteroplasmy in ovine fetuses and sheep cloned by somatic cell nuclear transfer. BMC Dev Biol 7:141. https://doi.org/10.1186/1471-213X-7-141

Cagnone G, Tsai T-SS, Srirattana K, Rossello F, Powell DR, Rohrer G, Cree L, Trounce IA, St John JC (2016a) Segregation of naturally occurring mitochondrial DNA variants in a mini-pig model. Genetics 202(3):931–944. https://doi.org/10.1534/genetics.115.181321

Cagnone GL, Tsai T-SS, Makanji Y, Matthews P, Gould J, Bonkowski MS, Elgass KD, Wong AS, Wu LE, McKenzie M, Sinclair DA, John JC (2016b) Restoration of normal embryogenesis by mitochondrial supplementation in pig oocytes exhibiting mitochondrial DNA deficiency. Sci Rep 6:23229. https://doi.org/10.1038/srep23229

Cao L, Shitara H, Horii T, Nagao Y, Imai H, Abe K, Hara T, Hayashi J-I, Yonekawa H (2007) The mitochondrial bottleneck occurs without reduction of mtDNA content in female mouse germ cells. Nat Genet 39(3):386–390. https://doi.org/10.1038/ng1970

Capaldi RA (1982) Arrangement of proteins in the mitochondrial inner membrane. Biochim Biophys Acta Rev Biomembr 694(3):291–306. https://doi.org/10.1016/0304-4157(82)90009-0

Castaneda CA, Kaye P, Pantaleon M, Phillips N, Norman S, Fry R, D'Occhio MJ (2013) Lipid content, active mitochondria and brilliant cresyl blue staining in bovine oocytes. Theriogenology 79(3):417–422. https://doi.org/10.1016/j.theriogenology.2012.10.006

Chang D, Clayton D (1985) Priming of human mitochondrial DNA replication occurs at the light-strand promoter. Proc Natl Acad Sci U S A 82(2):351–355. https://doi.org/10.1073/pnas.82.2.351

Chen D-Y, Wen D-C, Zhang Y-P, Sun Q-Y, Han Z-M, Liu Z-H, Shi P, Li J-S, Xiangyu J-G, Lian L, Kou Z-H, Wu Y-Q, Chen Y-C, Wang P-Y, Zhang H-M (2002) Interspecies implantation and mitochondria fate of panda-rabbit cloned embryos. Biol Reprod 67(2):637–642. https://doi.org/10.1095/biolreprod67.2.637

Cibelli JB, Campbell KH, Seidel GE, West MD, Lanza RP (2002) The health profile of cloned animals. Nat Biotechnol 20(1):13–14. https://doi.org/10.1038/nbt0102-13

Clay Montier LL, Deng J, Bai Y (2009) Number matters: control of mammalian mitochondrial DNA copy number. J Genet Genomics 36(3):125–131. https://doi.org/10.1016/S1673-8527(08)60099-5

Clayton DA (1982) Replication of animal mitochondrial DNA. Cell 28(4):693–705. https://doi.org/10.1016/0092-8674(82)90049-6

Clayton DA (2000) Transcription and replication of mitochondrial DNA. Hum Reprod 15(Suppl 2):11–17

Cohen J, Scott R, Schimmel T, Levron J, Willadsen S (1997) Birth of infant after transfer of anucleate donor oocyte cytoplasm into recipient eggs. Lancet 350(9072):186–187. https://doi.org/10.1016/S0140-6736(05)62353-7

Cortopassi GA, Shibata D, Soong NW, Arnheim N (1992) A pattern of accumulation of a somatic deletion of mitochondrial DNA in aging human tissues. Proc Natl Acad Sci U S A 89 (16):7370–7374

Cree L, Samuels D, de Sousa LS, Rajasimha H, Wonnapinij P, Mann J, Dahl H-HM, Chinnery P (2008) A reduction of mitochondrial DNA molecules during embryogenesis explains the rapid segregation of genotypes. Nat Genet 40(2):249–254. https://doi.org/10.1038/ng.2007.63

Cree LM, Hammond ER, Shelling AN, Berg MC, Peek JC, Green MP (2015) Maternal age and ovarian stimulation independently affect oocyte mtDNA copy number and cumulus cell gene expression in bovine clones. Hum Reprod 30(6):1410–1420. https://doi.org/10.1093/humrep/dev066

Danan C, Sternberg D, van Steirteghem A, Cazeneuve C, Duquesnoy P, Besmond C, Goossens M, Lissens W, Amselem S (1999) Evaluation of parental mitochondrial inheritance in neonates born after intracytoplasmic sperm injection. Am J Hum Genet 65(2):463–473

de Souza-Pinto NC, Mason PA, Hashiguchi K, Weissman L, Tian J, Guay D, Lebel M, Stevnsner TV, Rasmussen LJ, Bohr VA (2009) Novel DNA mismatch-repair activity involving YB-1 in human mitochondria. DNA Repair 8(6):704–719. https://doi.org/10.1016/j.dnarep.2009.01.021

Desjardins P, Frost E, Morais R (1985) Ethidium bromide-induced loss of mitochondrial DNA from primary chicken embryo fibroblasts. Mol Cell Biol 5(5):1163–1169

Dianov GL, Souza-Pinto N, Nyaga SG, Thybo T, Stevnsner T, Bohr VA (2001) Base excision repair in nuclear and mitochondrial DNA. Prog Nucleic Acid Res Mol Biol 68:285–297. https://doi.org/10.1016/S0079-6603(01)68107-8

Do M, Jang W-G, Hwang J, Jang H, Kim E-J, Jeong E-J, Shim H, Hwang S, Oh K, Byun S, Kim J-H, Lee J (2012) Inheritance of mitochondrial DNA in serially recloned pigs by somatic cell nuclear transfer (SCNT). Biochem Biophys Res Commun 424(4):765–770. https://doi.org/10.1016/j.bbrc.2012.07.031

Dokianakis E, Ladoukakis ED (2014) Different degree of paternal mtDNA leakage between male and female progeny in interspecific Drosophila crosses. Ecol Evol 4(13):2633–2641. https://doi.org/10.1002/ece3.1069

Douglas GC, VandeVoort CA, Kumar P, Chang T-C, Golos TG (2009) Trophoblast stem cells: models for investigating trophectoderm differentiation and placental development. Endocr Rev 30(3):228–240. https://doi.org/10.1210/er.2009-0001

Duarte Alcoba D, Gonsales Valério E, Conzatti M, Schneider J, Capp E, von Eye CH, Brum IS (2017) Selection of developmentally competent human oocytes aspirated during cesarean section. J Matern Fetal Neonatal Med:1–5. https://doi.org/10.1080/14767058.2017.1297405

El Shourbagy SH, Spikings EC, Freitas M, St. John JC (2006) Mitochondria directly influence fertilisation outcome in the pig. Reproduction 131(2):233–245. https://doi.org/10.1530/rep.1.00551

Ericsson SA, Boice ML, Funahashi H, Day BN (1993) Assessment of porcine oocytes using brilliant cresyl blue. Theriogenology 39(1):214. https://doi.org/10.1016/0093-691X(93)90069-H

Evans MJ, Gurer C, Loike JD, Wilmut I, Schnieke AE, Schon EA (1999) Mitochondrial DNA genotypes in nuclear transfer-derived cloned sheep. Nat Genet 23(1):90–93. https://doi.org/10.1038/12696

Facucho-Oliveira JM, St. John JC (2009) The relationship between pluripotency and mitochondrial DNA proliferation during early embryo development and embryonic stem cell differentiation. Stem Cell Rev Rep 5(2):140–158. https://doi.org/10.1007/s12015-009-9058-0

Facucho-Oliveira JM, Alderson J, Spikings EC, Egginton S, St. John JC (2007) Mitochondrial DNA replication during differentiation of murine embryonic stem cells. J Cell Sci 120 (22):4025–4034. https://doi.org/10.1242/jcs.016972

Fakruzzaman M, Bang J-I, Lee K-L, Kim S-S, Ha AN, Ghanem N, Han C-H, Cho K-W, White KL, Kong I-K (2013) Mitochondrial content and gene expression profiles in oocyte-derived embryos of cattle selected on the basis of brilliant cresyl blue staining. Anim Reprod Sci 142(1):19–27. https://doi.org/10.1016/j.anireprosci.2013.08.012

Fathi M, Ashry M, Salama A, Badr MR (2017) Developmental competence of Dromedary camel (Camelus dromedarius) oocytes selected using brilliant cresyl blue staining. Zygote 25:529–536. https://doi.org/10.1017/S0967199417000387

Ferreira CR, Meirelles FV, Yamazaki W, Chiaratti MR, Méo SC, Perecin F, Smith LC, Garcia JM (2007) The kinetics of donor cell mtDNA in embryonic and somatic donor cell-derived bovine embryos. Cloning Stem Cells 9(4):618–629. https://doi.org/10.1089/clo.2006.0082

Ferreira CR, Burgstaller JP, Perecin F, Garcia JM, Chiaratti MR, Méo SC, Müller M, Smith LC, Meirelles FV, Steinborn R (2010) Pronounced segregation of donor mitochondria introduced by bovine ooplasmic transfer to the female germ-line. Biol Reprod 82(3):563–571. https://doi.org/10.1095/biolreprod.109.080564

Finsterer J (2008) Leigh and Leigh-like syndrome in children and adults. Pediatr Neurol 39 (4):223–235. https://doi.org/10.1016/j.pediatrneurol.2008.07.013

Fish J, Raule N, Attardi G (2004) Discovery of a major D-loop replication origin reveals two modes of human mtDNA synthesis. Science 306(5704):2098

Folch J, Cocero MJ, Chesné P, Alabart JL, Domínguez V, Cognié Y, Roche A, Fernández-Árias A, Martí JI, Sánchez P, Echegoyen E, Beckers JF, Bonastre AS, Vignon X (2009) First birth of an animal from an extinct subspecies (Capra pyrenaica pyrenaica) by cloning. Theriogenology 71 (6):1026–1034. https://doi.org/10.1016/j.theriogenology.2008.11.005

Fu B, Ren L, Liu D, Ma J-Z, An T-Z, Yang X-Q, Ma H, Zhang D-J, Guo Z-H, Guo Y-Y, Zhu M, Bai J (2015) Subcellular characterization of porcine oocytes with different glucose-6-phosphate dehydrogenase activities. Asian Australas J Anim Sci 28(12):1703–1712. https://doi.org/10.5713/ajas.15.0051

Giles RE, Blanc H, Cann HM, Wallace DC (1980) Maternal inheritance of human mitochondrial DNA. Proc Natl Acad Sci U S A 77(11):6715–6719

Gómez MC, Pope CE, Giraldo A, Lyons LA, Harris RF, King AL, Cole A, Godke RA, Dresser BL (2004) Birth of African Wildcat cloned kittens born from domestic cats. Cloning Stem Cells 6 (3):247–258. https://doi.org/10.1089/clo.2004.6.247

Gómez MC, Pope CE, Kutner RH, Ricks DM, Lyons LA, Ruhe M, Dumas C, Lyons J, López M, Dresser BL, Reiser J (2008) Nuclear transfer of sand cat cells into enucleated domestic cat oocytes is affected by cryopreservation of donor cells. Cloning Stem Cells 10(4):469–483. https://doi.org/10.1089/clo.2008.0021

Goto Y-I, Nonaka I, Horai S (1990) A mutation in the tRNALeu(UUR) gene associated with the MELAS subgroup of mitochondrial encephalomyopathies. Nature 348(6302):651–653

Gray H, Wong TW (1992) Purification and identification of subunit structure of the human mitochondrial DNA polymerase. J Biol Chem 267(9):5835–5841

Grégoire M, Morais R, Quilliam MA, Gravel D (1984) On auxotrophy for pyrimidines of respiration-deficient chick embryo cells. Eur J Biochem 142(1):49–55. https://doi.org/10.1111/j.1432-1033.1984.tb08249.x

Gyllensten U, Wharton D, Josefsson A, Wilson AC (1991) Paternal inheritance of mitochondrial DNA in mice. Nature 352(6332):255–257

Hayakawa T, Noda M, Yasuda K, Yorifuji H, Taniguchi S, Miwa I, Sakura H, Terauchi Y, Hayashi J, Sharp GW, Kanazawa Y, Akanuma Y, Yazaki Y, Kadowaki T (1998) Ethidium bromide-induced inhibition of mitochondrial gene transcription suppresses glucose-stimulated insulin release in the mouse pancreatic beta-cell line betaHC9. J Biol Chem 273(32):20300–20307

Herrmann JM, Neupert W (2000) Protein transport into mitochondria. Curr Opin Microbiol 3 (2):210–214. https://doi.org/10.1016/S1369-5274(00)00077-1

Hiendleder S, Zakhartchenko V, Wenigerkind H, Reichenbach HD, Brüggerhoff K, Prelle K, Brem G, Stojkovic M, Wolf E (2003) Heteroplasmy in bovine fetuses produced by intra- and inter-subspecific somatic cell nuclear transfer: neutral segregation of nuclear donor mitochondrial DNA in various tissues and evidence for recipient cow mitochondria in fetal blood. Biol Reprod 68(1):159–166. https://doi.org/10.1095/biolreprod.102.008201

Holt IJ, Harding AE, Morgan-Hughes JA (1988) Deletions of muscle mitochondrial DNA in patients with mitochondrial myopathies. Nature 331(6158):717–719

Holt IJ, Lorimer HE, Jacobs HT (2000) Coupled leading- and lagging-strand synthesis of mammalian mitochondrial DNA. Cell 100(5):515–524. https://doi.org/10.1016/S0092-8674(00)80688-1

Houshmand M, Holme E, Hanson C, Wennerholm U-B, Hamberger L (1997) Is paternal mitochondrial DNA transferred to the offspring following intracytoplasmic sperm injection? J Assist Reprod Genet 14(4):223–227. https://doi.org/10.1007/BF02766114

Hua S, Zhang Y, Li XC, Ma LB, Cao JW, Dai JP, Li R (2007) Effects of granulosa cell mitochondria transfer on the early development of bovine embryos in vitro. Cloning Stem Cells 9(2):237–246. https://doi.org/10.1089/clo.2006.0020

Hwang I, Jeong Y, Kim J, Lee H, Kang M, Park K, Park J, Kim Y, Kim W, Shin T, Hyun S, Jeung E-B, Hwang W (2013) Successful cloning of coyotes through interspecies somatic cell nuclear transfer using domestic dog oocytes. Reprod Fertil Dev 25(8):1142–1148. https://doi.org/10.1071/RD12256

Imsoonthornruksa S, Srirattana K, Phewsoi W, Tunwattana W, Parnpai R, Ketudat-Cairns M (2012) Segregation of donor cell mitochondrial DNA in gaur-bovine interspecies somatic cell nuclear transfer embryos, fetuses and an offspring. Mitochondrion 12(5):506–513. https://doi.org/10.1016/j.mito.2012.07.108

Ingman M, Kaessmann H, Paabo S, Gyllensten U (2000) Mitochondrial genome variation and the origin of modern humans. Nature 408(6813):708–713

Inoue K, Ogonuki N, Yamamoto Y, Takano K, Miki H, Mochida K, Ogura A (2004) Tissue-specific distribution of donor mitochondrial DNA in cloned mice produced by somatic cell nuclear transfer. Genesis (New York: 2000) 39(2):79–83. https://doi.org/10.1002/gene.20029

Iwata H, Goto H, Tanaka H, Sakaguchi Y, Kimura K, Kuwayama T, Monji Y (2011) Effect of maternal age on mitochondrial DNA copy number, ATP content and IVF outcome of bovine oocytes. Reprod Fertil Dev 23(3):424–432. https://doi.org/10.1071/RD10133

Jenuth JP, Peterson AC, Fu K, Shoubridge EA (1996) Random genetic drift in the female germline explains the rapid segregation of mammalian mitochondrial DNA. Nat Genet 14(2):146–151. https://doi.org/10.1038/ng1096-146

Jeppesen TD, Schwartz M, Frederiksen AL, Wibrand F, Olsen DB, Vissing J (2006) Muscle phenotype and mutation load in 51 persons with the 3243A>G mitochondrial DNA mutation. Arch Neurol 63(12):1701–1706. https://doi.org/10.1001/archneur.63.12.1701

Jiang Y, Kelly R, Peters A, Fulka H, Dickinson A, Mitchell DA, St. John JC (2011) Interspecies somatic cell nuclear transfer is dependent on compatible mitochondrial DNA and reprogramming factors. PLoS One 6(4):e14805. https://doi.org/10.1371/journal.pone.0014805

Keilbaugh SA, Hobbs GA, Simpson MV (1993) Anti-human immunodeficiency virus type 1 therapy and peripheral neuropathy: prevention of 2′,3′-dideoxycytidine toxicity in PC12 cells, a neuronal model, by uridine and pyruvate. Mol Pharmacol 44(4):702

Kelly RDW, St. John JC (2011) Role of mitochondrial DNA replication during differentiation of reprogrammed stem cells. Int J Dev Biol 54(11–12):1659–1670. https://doi.org/10.1387/ijdb.103202rk

Kelly R, Mahmud A, McKenzie M, Trounce I, St John J (2012) Mitochondrial DNA copy number is regulated in a tissue specific manner by DNA methylation of the nuclear-encoded DNA polymerase gamma A. Nucleic Acids Res 40(20):10124–10138. https://doi.org/10.1093/nar/gks770

Kelly RDW, Sumer H, McKenzie M, Facucho-Oliveira J, Trounce IA, Verma PJ, St. John JC (2013) The effects of nuclear reprogramming on mitochondrial DNA replication. Stem Cell Rev Rep 9(1):1–15. https://doi.org/10.1007/s12015-011-9318-7

Kim M, Jang G, Oh H, Yuda F, Kim H, Hwang W, Hossein M, Kim J, Shin N, Kang S, Lee B (2007) Endangered wolves cloned from adult somatic cells. Cloning Stem Cells 9(1):130–137. https://doi.org/10.1089/clo.2006.0034

King M, Attardi G (1989) Human cells lacking mtDNA: repopulation with exogenous mitochondria by complementation. Science (New York, NY) 246(4929):500–503. https://doi.org/10.1126/science.2814477

Korhonen JA, Gaspari M, Falkenberg M (2003) TWINKLE Has 5′ → 3′ DNA helicase activity and is specifically stimulated by mitochondrial single-stranded DNA-binding protein. J Biol Chem 278(49):48627–48632. https://doi.org/10.1074/jbc.M306981200

Krishnan KJ, Greaves LC, Reeve AK, Turnbull D (2007) The ageing mitochondrial genome. Nucleic Acids Res 35(22):7399–7405. https://doi.org/10.1093/nar/gkm635

Kujoth GC, Hiona A, Pugh TD, Someya S, Panzer K, Wohlgemuth SE, Hofer T, Seo AY, Sullivan R, Jobling WA, Morrow JD, van Remmen H, Sedivy JM, Yamasoba T, Tanokura M, Weindruch R, Leeuwenburgh C, Prolla TA (2005) Mitochondrial DNA mutations, oxidative stress, and apoptosis in mammalian aging. Science 309(5733):481

Lanza R, Cibelli J, Diaz F, Moraes C, Farin P, Farin C, Hammer C, West M, Damiani P (2000) Cloning of an endangered species (Bos gaurus) using interspecies nuclear transfer. Cloning 2 (2):79–90. https://doi.org/10.1089/152045500436104

Larsson N-G (2010) Somatic mitochondrial DNA mutations in mammalian aging. Annu Rev Biochem 79(1):683–706. https://doi.org/10.1146/annurev-biochem-060408-093701

Larsson NG, Tulinius MH, Holme E, Oldfors A, Andersen O, Wahlström J, Aasly J (1992) Segregation and manifestations of the mtDNA tRNA(Lys) A-->G(8344) mutation of myoclonus epilepsy and ragged-red fibers (MERRF) syndrome. Am J Hum Genet 51(6):1201–1212

Lee WT, St John J (2015) The control of mitochondrial DNA replication during development and tumorigenesis. Ann N Y Acad Sci 1350:95–106. https://doi.org/10.1111/nyas.12873

Lee J-H, Peters A, Fisher P, Bowles E, St John J, Campbell K (2010a) Generation of mtDNA homoplasmic cloned lambs. Cell Reprogram 12(3):347–355. https://doi.org/10.1089/cell.2009.0096

Lee JH, Peters A, Fisher P, Bowles EJ, St. John JC, Campbell KHS (2010b) Generation of mtDNA homoplasmic cloned lambs. Cell Reprogram 12(3):347–355. https://doi.org/10.1089/cell.2009.0096

Li K, Li Y, Shelton JM, Richardson JA, Spencer E, Chen ZJ, Wang X, Williams RS (2000) Cytochrome c deficiency causes embryonic lethality and attenuates stress-induced apoptosis. Cell 101(4):389–399. https://doi.org/10.1016/S0092-8674(00)80849-1

Liang C-G, Han Z, Cheng Y, Zhong Z, Latham KE (2009) Effects of ooplasm transfer on paternal genome function in mice. Hum Reprod 24(11):2718–2728. https://doi.org/10.1093/humrep/dep286

Lloyd RE, Lee JH, Alberio R, Bowles EJ, Ramalho-Santos J, Campbell KHS, St. John JC (2006) Aberrant nucleo-cytoplasmic cross-talk results in donor cell mtDNA persistence in cloned embryos. Genetics 172(4):2515–2527. https://doi.org/10.1534/genetics.105.055145

Loi P, Ptak G, Barboni B, Fulka J, Cappai P, Clinton M (2001) Genetic rescue of an endangered mammal by cross-species nuclear transfer using post-mortem somatic cells. Nat Biotechnol 19 (10):962–964. https://doi.org/10.1038/nbt1001-962

Maceluch J, Niedziela M (2006) The clinical diagnosis and molecular genetics of Kearns-Sayre syndrome: a complex mitochondrial encephalomyopathy. Pediatr Endocrinol Rev 4(2):117–137

Man PYW, Turnbull DM, Chinnery PF (2002) Leber hereditary optic neuropathy. J Med Genet 39 (3):162

Mangia F, Epstein CJ (1975) Biochemical studies of growing mouse oocytes: preparation of oocytes and analysis of glucose-6-phosphate dehydrogenase and lactate dehydrogenase activities. Dev Biol 45(2):211–220. https://doi.org/10.1016/0012-1606(75)90061-5

Manjunatha BM, Gupta PSP, Devaraj M, Ravindra JP, Nandi S (2007) Selection of developmentally competent buffalo oocytes by brilliant cresyl blue staining before IVM. Theriogenology 68 (9):1299–1304. https://doi.org/10.1016/j.theriogenology.2007.08.031

Marchington DR, Scott Brown MSG, Lamb VK, van Golde RJT, Kremer JAM, Tuerlings JHAM, Mariman ECM, Balen AH, Poulton J (2002) No evidence for paternal mtDNA transmission to offspring or extra-embryonic tissues after ICSI. Mol Hum Reprod 8(11):1046–1049

Mastromonaco GF, King WA (2007) Cloning in companion animal, non-domestic and endangered species: can the technology become a practical reality? Reprod Fertil Dev 19(6):748–761

May-Panloup P, Chrétien MF, Jacques C, Vasseur C, Malthièry Y, Reynier P (2005a) Low oocyte mitochondrial DNA content in ovarian insufficiency. Hum Reprod 20(3):593–597. https://doi.org/10.1093/humrep/deh667

May-Panloup P, Vignon X, Chrétien M-FF, Heyman Y, Tamassia M, Malthièry Y, Reynier P (2005b) Increase of mitochondrial DNA content and transcripts in early bovine embryogenesis associated with upregulation of mtTFA and NRF1 transcription factors. Reprod Biol Endocrinol 3:65. https://doi.org/10.1186/1477-7827-3-65

McFarland R, Taylor RW, Turnbull DM (2007) Mitochondrial disease—its impact, etiology, and pathology. Curr Top Dev Biol 77:113–155. https://doi.org/10.1016/S0070-2153(06)77005-3

McKenzie M, Trounce I (2000) Expression of Rattus norvegicus mtDNA in Mus musculus cells results in multiple respiratory chain defects. J Biol Chem 275(40):31514–31519. https://doi.org/10.1074/jbc.M004070200

Meirelles FV, Bordignon V, Watanabe Y, Watanabe M, Dayan A, Lôbo RB, Garcia JM, Smith LC (2001) Complete replacement of the mitochondrial genotype in a Bos indicus calf reconstructed by nuclear transfer to a Bos taurus oocyte. Genetics 158(1):351

Mohammadi-Sangcheshmeh A, Held E, Ghanem N, Rings F, Salilew-Wondim D, Tesfaye D, Sieme H, Schellander K, Hoelker M (2011) G6PDH-activity in equine oocytes correlates with morphology, expression of candidate genes for viability, and preimplantative in vitro development. Theriogenology 76(7):1215–1226. https://doi.org/10.1016/j.theriogenology.2011.05.025

Mohammadi-Sangcheshmeh A, Soleimani M, Deldar H, Salehi M, Soudi S, Hashemi SM, Schellander K, Hoelker M (2012) Prediction of oocyte developmental competence in ovine using glucose-6-phosphate dehydrogenase (G6PDH) activity determined at retrieval time. J Assist Reprod Genet 29(2):153–158. https://doi.org/10.1007/s10815-011-9625-6

Moraes CT, DiMauro S, Zeviani M, Lombes A, Shanske S, Miranda AF, Nakase H, Bonilla E, Werneck LC, Servidei S, Nonaka I, Koga Y, Spiro AJ, Brownell AKW, Schmidt B, Schotland DL, Zupanc M, DeVivo DC, Schon EA, Rowland LP (1989) Mitochondrial DNA deletions in progressive external ophthalmoplegia and Kearns-Sayre syndrome. N Engl J Med 320 (20):1293–1299. https://doi.org/10.1056/NEJM198905183202001

Moyes CD, Battersby BJ, Leary SC (1998) Regulation of muscle mitochondrial design. J Exp Biol 201(3):299

Nakada K, Inoue K, Ono T, Isobe K, Ogura A, Goto YI, Nonaka I, Hayashi JI (2001) Inter-mitochondrial complementation: mitochondria-specific system preventing mice from expression of disease phenotypes by mutant mtDNA. Nat Med 7(8):934–940. https://doi.org/10.1038/90976

Nelson I, Hanna MG, Wood NW, Harding AE (1997) Depletion of mitochondrial DNA by ddC in untransformed human cell lines. Somat Cell Mol Genet 23(4):287–290. https://doi.org/10.1007/BF02674419

Palermo G, Joris H, Devroey P, van Steirteghem AC (1992) Pregnancies after intracytoplasmic injection of single spermatozoon into an oocyte. Lancet 340(8810):17–18. https://doi.org/10.1016/0140-6736(92)92425-F

Pfeiffer T, Schuster S, Bonhoeffer S (2001) Cooperation and competition in the evolution of ATP-producing pathways. Science (New York, NY) 292(5516):504–507. https://doi.org/10.1126/science.1058079

Pinkert CA, Irwin MH, Johnson LW, Moffatt RJ (1997) Mitochondria transfer into mouse ova by microinjection. Transgenic Res 6(6):379–383

Pujol M, López-Béjar M, Paramio MT (2004) Developmental competence of heifer oocytes selected using the brilliant cresyl blue (BCB) test. Theriogenology 61(4):735–744. https://doi.org/10.1016/S0093-691X(03)00250-4

Rahman S, Poulton J (2008) Diagnosis of mitochondrial DNA depletion syndromes. Arch Dis Child 94(1):3

Ramos A, Santos C, Mateiu L, Gonzalez MM, Alvarez L, Azevedo L, Amorim A, Aluja MP (2013) Frequency and pattern of heteroplasmy in the complete human mitochondrial genome. PLoS One 8(10):e74636. https://doi.org/10.1371/journal.pone.0074636

Reynier P, May-Panloup P, Chrétien MF, Morgan CJ, Jean M, Savagner F, Barrière P, Malthièry Y (2001) Mitochondrial DNA content affects the fertilizability of human oocytes. Mol Hum Reprod 7(5):425–429

Rodrigues BA, Rodriguez P, Silva AEF, Cavalcante LF, Feltrin C, Rodrigues JL (2009) Preliminary study in immature canine oocytes stained with brilliant cresyl blue and obtained from bitches with low and high progesterone serum profiles. Reprod Domest Anim 44:255–258. https://doi.org/10.1111/j.1439-0531.2009.01408.x

Rötig A, Cormier V, Blanche S, Bonnefont JP, Ledeist F, Romero N, Schmitz J, Rustin P, Fischer A, Saudubray JM (1990) Pearson's marrow-pancreas syndrome. A multisystem mitochondrial disorder in infancy. J Clin Invest 86(5):1601–1608

Salviano MB, Collares FJF, Becker BS, Rodrigues BA, Rodrigues JL (2015) Bovine non-competent oocytes (BCB–) negatively impact the capacity of competent (BCB+) oocytes to undergo in vitro maturation, fertilisation and embryonic development. Zygote 24 (2):245–251. https://doi.org/10.1017/S0967199415000118

Santos TA, El Shourbagy S, St. John JC (2006) Mitochondrial content reflects oocyte variability and fertilization outcome. Fertil Steril 85(3):584–591. https://doi.org/10.1016/j.fertnstert.2005.09.017

Schriner SE, Linford NJ, Martin GM, Treuting P, Ogburn CE, Emond M, Coskun PE, Ladiges W, Wolf N, van Remmen H, Wallace DC, Rabinovitch PS (2005) Extension of murine life span by overexpression of catalase targeted to mitochondria. Science 308(5730):1909

Sekirina GG, Bogoliubova NA, Antonova NV, Dyban AP (1997) The behaviour of mitochondria and cell integration during somatic hybridisation of sister blastomeres of the 2-cell mouse embryo. Zygote 5(2):97–103

Shadel GS, Clayton DA (1997) Mitochondrial DNA maintenance in vertebrates. Annu Rev Biochem 66(1):409–435. https://doi.org/10.1146/annurev.biochem.66.1.409

Sharpley MS, Marciniak C, Eckel-Mahan K, McManus M, Crimi M, Waymire K, Lin CS, Masubuchi S, Friend N, Koike M, Chalkia D, MacGregor G, Sassone-Corsi P, Wallace DC (2012) Heteroplasmy of mouse mtDNA is genetically unstable and results in altered behavior and cognition. Cell 151(2):333–343. https://doi.org/10.1016/j.cell.2012.09.004

Shoffner JM, Wallace DC (1990) Oxidative phosphorylation diseases. Disorders of two genomes. Adv Hum Genet 19:267–332

Shoffner JM, Lott MT, Lezza AMS, Seibel P, Ballinger SW, Wallace DC (1990) Myoclonic epilepsy and ragged-red fiber disease (MERRF) is associated with a mitochondrial DNA tRNALys mutation. Cell 61(6):931–937. https://doi.org/10.1016/0092-8674(90)90059-N

Shoubridge EA (2000) Mitochondrial DNA segregation in the developing embryo. Hum Reprod 15 (Suppl 2):229–234

Shoubridge E, Wai T (2007) Mitochondrial DNA and the mammalian oocyte. Curr Top Dev Biol 77:87–111. https://doi.org/10.1016/S0070-2153(06)77004-1

Silva C, Sartorelli E, Castilho A, Satrapa R, Puelker R, Razza E, Ticianelli J, Eduardo H, Loureiro B, Barros C (2013) Effects of heat stress on development, quality and survival of Bos indicus and Bos taurus embryos produced in vitro. Theriogenology 79(2):351–357. https://doi.org/10.1016/j.theriogenology.2012.10.003

Spikings E, Alderson J, St John J (2006) Transmission of mitochondrial DNA following assisted reproduction and nuclear transfer. Hum Reprod Update 12(4):401–415. https://doi.org/10.1093/humupd/dml011

Spikings EC, Alderson J, St. John JC (2007) Regulated mitochondrial DNA replication during oocyte maturation is essential for successful porcine embryonic development. Biol Reprod 76 (2):327–335. https://doi.org/10.1095/biolreprod.106.054536

Srirattana K, St. John JC (2017) Manipulating the mitochondrial genome to enhance cattle embryo development. G3 7(7):2065

Srirattana K, St. John JC (2018) Additional mitochondrial DNA influences the interactions between the nuclear and mitochondrial genomes in a bovine embryo model of nuclear transfer. Sci Rep 8 (1):7246. https://doi.org/10.1038/s41598-018-25516-3

Srirattana K, Matsukawa K, Akagi S, Tasai M, Tagami T, Nirasawa K, Nagai T, Kanai Y, Parnpai R, Takeda K (2011) Constant transmission of mitochondrial DNA in intergeneric

cloned embryos reconstructed from swamp buffalo fibroblasts and bovine ooplasm. Anim Sci J 82(2):236–243. https://doi.org/10.1111/j.1740-0929.2010.00827.x
Srirattana K, Imsoonthornruksa S, Laowtammathron C, Sangmalee A, Tunwattana W, Thongprapai T, Chaimongkol C, Ketudat-Cairns M, Parnpai R (2012) Full-term development of gaur-bovine interspecies somatic cell nuclear transfer embryos: effect of trichostatin A treatment. Cell Reprogram 14(3):248–257. https://doi.org/10.1089/cell.2011.0099
St John J (2012) Transmission, inheritance and replication of mitochondrial DNA in mammals: implications for reproductive processes and infertility. Cell Tissue Res 349(3):795–808. https://doi.org/10.1007/s00441-012-1444-2
St. John JC, Moffatt O, D'Souza N (2005) Aberrant heteroplasmic transmission of mtDNA in cloned pigs arising from double nuclear transfer. Mol Reprod Dev 72(4):450–460. https://doi.org/10.1002/mrd.20370
St. John JC, Facucho-Oliveira J, Jiang Y, Kelly R, Salah R (2010) Mitochondrial DNA transmission, replication and inheritance: a journey from the gamete through the embryo and into offspring and embryonic stem cells. Hum Reprod Update 16(5):488–509. https://doi.org/10.1093/humupd/dmq002
Steinborn R, Schinogl P, Zakhartchenko V, Achmann R, Schernthaner W, Stojkovic M, Wolf E, Müller M, Brem G (2000) Mitochondrial DNA heteroplasmy in cloned cattle produced by fetal and adult cell cloning. Nat Genet 25(3):255–257. https://doi.org/10.1038/77000
Steinborn R, Schinogl P, Wells DN, Bergthaler A, Müller M, Brem G (2002) Coexistence of Bos taurus and B. indicus mitochondrial DNAs in nuclear transfer-derived somatic cattle clones. Genetics 162(2):823–829
Su J, Wang Y, Li R, Peng H, Hua S, Li Q, Quan F, Guo Z, Zhang Y (2012) Oocytes selected using BCB staining enhance nuclear reprogramming and the in vivo development of SCNT embryos in cattle. PLoS One 7(4):e36181. https://doi.org/10.1371/journal.pone.0036181
Sugulle AH, Dochi O, Koyama H (2008) Developmental competence of bovine oocytes selected by Brilliant Cresyl Blue Staining: effect of the presence of corpus luteum on embryo development. J Mamm Ova Res 25(1):50–55. https://doi.org/10.1274/jmor.25.50
Sun X, St. John JC (2016) The role of the mtDNA set point in differentiation, development and tumorigenesis. Biochem J 473(19):2955
Sutovsky P, Moreno R, Ramalho-Santos J, Dominko T, Simerly C, Schatten G (1999) Ubiquitin tag for sperm mitochondria. Nature 402(6760):371–372. https://doi.org/10.1038/46466
Tachibana M, Sparman M, Sritanaudomchai H, Ma H, Clepper L, Woodward J, Li Y, Ramsey C, Kolotushkina O, Mitalipov S (2009) Mitochondrial gene replacement in primate offspring and embryonic stem cells. Nature 461(7262):367–372. https://doi.org/10.1038/nature08368
Tait SWG, Green DR (2010) Mitochondria and cell death: outer membrane permeabilization and beyond. Nat Rev Mol Cell Biol 11:621–632
Takeda K, Takahashi S, Onishi A, Goto Y, Miyazawa A, Imai H (1999) Dominant distribution of mitochondrial DNA from recipient oocytes in bovine embryos and offspring after nuclear transfer. J Reprod Fertil 116(2):253–259. https://doi.org/10.1530/jrf.0.1160253
Takeda K, Akagi S, Kaneyama K, Kojima T, Takahashi S, Imai H, Yamanaka M, Onishi A, Hanada H (2003) Proliferation of donor mitochondrial DNA in nuclear transfer calves (Bos taurus) derived from cumulus cells. Mol Reprod Dev 64(4):429–437. https://doi.org/10.1002/mrd.10279
Takeda K, Tasai M, Iwamoto M, Onishi A, Tagami T, Nirasawa K, Hanada H, Pinkert CA (2005) Microinjection of cytoplasm or mitochondria derived from somatic cells affects parthenogenetic development of murine oocytes. Biol Reprod 72(6):1397–1404. https://doi.org/10.1095/biolreprod.104.036129
Takeda K, Tasai M, Iwamoto M, Akita T, Tagami T, Nirasawa K, Hanada H, Onishi A (2006) Transmission of mitochondrial DNA in pigs and progeny derived from nuclear transfer of Meishan pig fibroblast cells. Mol Reprod Dev 73(3):306–312. https://doi.org/10.1002/mrd.20403

Takeda K, Tasai M, Akagi S, Matsukawa K, Takahashi S, Iwamoto M, Srirattana K, Onishi A, Tagami T, Nirasawa K, Hanada H, Pinkert CA (2010) Microinjection of serum-starved mitochondria derived from somatic cells affects parthenogenetic development of bovine and murine oocytes. Mitochondrion 10(2):137–142. https://doi.org/10.1016/j.mito.2009.12.144

Takeo S, Goto H, Kuwayama T, Monji Y, Iwata H (2013) Effect of maternal age on the ratio of cleavage and mitochondrial DNA copy number in early developmental stage bovine embryos. J Reprod Dev 59(2):174–179. https://doi.org/10.1262/jrd.2012-148

Thongphakdee A, Kobayashi S, Imai K, Inaba Y, Tasai M, Tagami T, Nirasawa K, Nagai T, Saito N, Techakumphu M, Takeda K (2008) Interspecies nuclear transfer embryos reconstructed from cat somatic cells and bovine ooplasm. J Reprod Dev 54(2):142–147. https://doi.org/10.1262/jrd.19159

Thundathil J, Filion F, Smith LC (2005) Molecular control of mitochondrial function in preimplantation mouse embryos. Mol Reprod Dev 71(4):405–413. https://doi.org/10.1002/mrd.20260

Torroni A, Lott MT, Cabell MF, Chen YS, Lavergne L, Wallace DC (1994) mtDNA and the origin of Caucasians: identification of ancient Caucasian-specific haplogroups, one of which is prone to a recurrent somatic duplication in the D-loop region. Am J Hum Genet 55(4):760–776

Trifunovic A, Wredenberg A, Falkenberg M, Spelbrink JN, Rovio AT, Bruder CE, Bohlooly-Y M, Gidlof S, Oldfors A, Wibom R, Tornell J, Jacobs HT, Larsson N-G (2004) Premature ageing in mice expressing defective mitochondrial DNA polymerase. Nature 429(6990):417–423. http://www.nature.com/nature/journal/v429/n6990/suppinfo/nature02517_S1.html

Tzeng CR, Hsieh RH, Au HK, Yen YH, Chang SJ, Cheng YF (2004) Mitochondria transfer (MIT) into oocyte from autologous cumulus granulosa cells (cGCs). Fertil Steril 82:S53. https://doi.org/10.1016/j.fertnstert.2004.07.136

Upholt WB, Dawid IB (1977) Mapping of mitochondrial DNA of individual sheep and goats: rapid evolution in the D loop region. Cell 11(3):571–583. https://doi.org/10.1016/0092-8674(77)90075-7

Ursing BM, Arnason U (1998) The complete mitochondrial DNA sequence of the pig (Sus scrofa). J Mol Evol 47(3):302–306. https://doi.org/10.1007/PL00006388

Vakifahmetoglu-Norberg H, Ouchida AT, Norberg E (2017) The role of mitochondria in metabolism and cell death. Biochem Biophys Res Commun 482(3):426–431. https://doi.org/10.1016/j.bbrc.2016.11.088

van Blerkom J, Davis PW, Lee J (1995) Fertilization and early embryolgoy: ATP content of human oocytes and developmental potential and outcome after in-vitro fertilization and embryo transfer. Hum Reprod 10(2):415–424

van Blerkom J, Sinclair J, Davis P (1998) Mitochondrial transfer between oocytes: potential applications of mitochondrial donation and the issue of heteroplasmy. Hum Reprod 13(10):2857–2868

Wai T, Teoli D, Shoubridge E (2008) The mitochondrial DNA genetic bottleneck results from replication of a subpopulation of genomes. Nat Genet 40(12):1484–1488. https://doi.org/10.1038/ng.258

Wai T, Ao A, Zhang X, Cyr D, Dufort D, Shoubridge EA (2010) The role of mitochondrial DNA copy number in mammalian fertility. Biol Reprod 83(1):52–62. https://doi.org/10.1095/biolreprod.109.080887

Wakefield SL, Lane M, Mitchell M (2011) Impaired mitochondrial function in the preimplantation embryo perturbs fetal and placental development in the mouse. Biol Reprod 84(3):572–580. https://doi.org/10.1095/biolreprod.110.087262

Wallace DC, Singh G, Lott MT, Hodge JA, Schurr TG, Lezza AM, Elsas LJ, Nikoskelainen EK (1988) Mitochondrial DNA mutation associated with Leber's hereditary optic neuropathy. Science 242(4884):1427

Wang L, Lin J, Huang J, Wang J, Zhao Y, Chen T (2012) Selection of ovine oocytes by brilliant cresyl blue staining. J Biomed Biotechnol 2012:7. https://doi.org/10.1155/2012/161372

Waring MJ (1965) Complex formation between ethidium bromide and nucleic acids. J Mol Biol 13(1):269–282. https://doi.org/10.1016/S0022-2836(65)80096-1

Wilmut I, Schnieke AE, McWhir J, Kind AJ, Campbell KHS (1997) Viable offspring derived from fetal and adult mammalian cells. Nature 385(6619):810–813. https://doi.org/10.1038/385810a0
Wong L-JC (2007) Diagnostic challenges of mitochondrial DNA disorders. Mitochondrion 7 (1–2):45–52. https://doi.org/10.1016/j.mito.2006.11.025
Wu Y-G, Liu Y, Zhou P, Lan G-C, Han D, Miao D-Q, Tan J-H (2007) Selection of oocytes for in vitro maturation by brilliant cresyl blue staining: a study using the mouse model. Cell Res 17 (8):722–731
Yi YC, Chen MJ, Ho JYP, Guu HF, Ho ESC (2007) Mitochondria transfer can enhance the murine embryo development. J Assist Reprod Genet 24(10):445–449. https://doi.org/10.1007/s10815-007-9161-6
Yoneda M, Tanno Y, Horai S, Ozawa T, Miyatake T, Tsuji S (1990) A common mitochondrial DNA mutation in the t-RNA(Lys) of patients with myoclonus epilepsy associated with ragged-red fibers. Biochem Int 21(5):789–796
Zhang J, Liu H, Luo S, Lu Z, Chávez-Badiola A, Liu Z, Yang M, Merhi Z, Silber SJ, Munné S, Konstandinidis M, Wells D, Tan JJ, Huang T (2017) Live birth derived from oocyte spindle transfer to prevent mitochondrial disease. Reprod Biomed Online 34(4):361–368. https://doi.org/10.1016/j.rbmo.2017.01.013
Zhao X, Li N, Guo W, Hu X, Liu Z, Gong G, Wang A, Feng J, Wu C (2004) Further evidence for paternal inheritance of mitochondrial DNA in the sheep (Ovis aries). Heredity 93(4):399–403
Zhao N, Zhang Y, Liu Q, Xiang W (2015) Mfn2 affects embryo development via mitochondrial dysfunction and apoptosis. PLoS One 10(5):e0125680. https://doi.org/10.1371/journal.pone.0125680
Zouros E, Freeman KR, Ball AO, Pogson GH (1992) Direct evidence for extensive paternal mitochondrial DNA inheritance in the marine mussel Mytilus. Nature 359:412. https://doi.org/10.1038/359412a0

Mitochondria Inspire a Lifestyle

Peter Kramer and Paola Bressan

Now, here, you see, it takes all the running you can do, to keep in the same place.
—Lewis Carroll, Through the Looking-Glass and What Alice Found There

Contents

Abstract Tucked inside our cells, we animals (and plants, and fungi) carry mitochondria, minuscule descendants of bacteria that invaded our common ancestor 2 billion years ago. This unplanned breakthrough endowed our ancestors with a convenient, portable source of energy, enabling them to progress towards more ambitious forms of life. Mitochondria still manufacture most of our energy; we have evolved to invest it to grow and produce offspring, and to last long enough to make it all happen. Yet because the continuous generation of energy is inevitably linked to that of toxic free radicals, mitochondria give us life and give us death. Stripping away clutter and minutiae, here we present a big-picture perspective of how mitochondria work, how they are passed on virtually only by mothers, and how

P. Kramer and P. Bressan (✉)
Dipartimento di Psicologia Generale, University of Padova, Padova, Italy
e-mail: peter.kramer@unipd.it; paola.bressan@unipd.it

P. Sutovsky (ed.), *Cellular and Molecular Basis of Mitochondrial Inheritance*, Advances in Anatomy, Embryology and Cell Biology 231,
https://doi.org/10.1007/102_2018_5, Published online: 5 January 2019

they shape the lifestyles of species and individuals. We discuss why restricting food prolongs lifespan, why reproducing shortens it, and why moving about protects us from free radicals despite increasing their production. We show that our immune cells use special mitochondria to keep control over our gut microbes. And we lay out how the fabrication of energy and free radicals sets the internal clocks that command our everyday rhythms—waking, eating, sleeping. Mitochondria run the show.

Keywords Antioxidants, Circadian rhythms, Dietary restriction, Free radicals, Longevity, Mitochondria

1 Pump or Perish

Mitochondria were once bacteria that invaded one-cell creatures from another domain of life, and that happened to let out as waste precisely those gasses that the host creatures used to sustain themselves (Martin et al. 2016). The cohabitation went spectacularly well and the guest bacteria ended up relinquishing part of their genetic material to the host cell—which out of it built a control center, a "nucleus"—while keeping only a few genes to themselves. This unplanned breakthrough endowed the hosts with a convenient, portable source of energy, making it possible for them to evolve into more ambitious forms of life (of which we are one).

From their bacterial ancestor, mitochondria have inherited the ability to produce energy via an intricate mechanism whose details differ little across the various kingdoms. Digested food is burned to fuel a machine that, from ADP (adenosine *di*phosphate) and an extra phosphate, forges ATP (adenosine *tri*phosphate). The freshly coined ATP is then dispatched from the mitochondrion to the host cell. Wherever the cell needs it, energy is released from ATP by stripping it of a phosphate. This leaves behind ADP and phosphate, that within the mitochondrion can be recharged to ATP again. What is exquisitely elaborate is how mitochondria put the plan into effect.

Each mitochondrion features a fairly smooth outer membrane, that separates the mitochondrion from the rest of the host cell, and a much more convoluted and interesting inner membrane; together they enclose the intermembrane space (Fig. 1). The inner membrane is important enough that an average human carries 14,000 m^2 of it (Rich 2003). Plugged through the membrane are *complexes*, minuscule pieces of machinery dedicated to the job of ATP generation; a single mammalian mitochondrion accommodates legions of them, perhaps in the order of 30,000 (Lane 2005). These are organized into sets of four consecutive stations. In the space in the middle of the mitochondrion, the matrix, equal numbers of protons and electrons are stripped from the molecular remains of digested food; the electrons are separated from the protons and fed into the first machine, called complex I. From there, the electrons are shuttled along the inner membrane through two further stations[1] and finally disgorged back into the matrix, where they are sucked up by oxygen.

[1]Only two, because one of the four stations is just an alternative entry point for electrons.

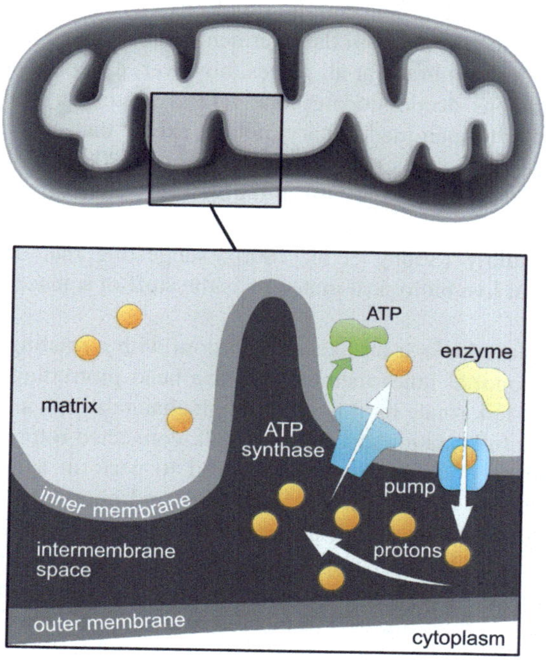

Fig. 1 Schematic, simplified representation of a mitochondrion. Inset: in the matrix (light gray), an enzyme (yellow) delivers protons (orange) and electrons stripped from digested food to a proton pump (blue shape on the right; electrons and additional stations are not shown). The flow of electrons within the pump provides the energy to make it transfer protons from the matrix to the intermembrane space (dark gray). The protons can then return from the intermembrane space to the matrix via the ATP synthase (mushroom-like blue shape on the left). In the process, the ATP synthase is pressed into service and produces ATP (green). Image source: Bressan (2018)

The passage of electrons *along* the membrane only serves to power the passage of protons *across* it—which is indeed the whole point of the arrangement. Making its way within each station, an electron creates disturbances that end up opening, one after the other, a series of channels across the membrane. Each channel stays open just long enough to let a proton pass through and shuts closed right behind it. The end result is that the protons are transferred from one side (the matrix) to the other (the intermembrane space) of a membrane that is otherwise sealed off to them. For every pair of electrons that reaches the end of the chain to be snatched up by oxygen, ten protons have been pumped across the membrane.

Each proton carries one positive charge. Thus, the amassing of protons on one side of the membrane produces a large difference in electrical charge, with the outside (the intermembrane space) becoming strongly positive relative to the inside (the matrix). The voltage difference across the membrane amounts to one-tenth of that of an alkaline watch battery; yet because the membrane is extraordinarily thin, we are talking about an electric field strength in the order of 30 million volts per meter, much like a bolt of lightning (Lane 2005). As the number of protons builds up, the difference in proton concentration and electrical charge between the two sides tends to pull the protons from the intermembrane space to the matrix. Protons are held back by the membrane at all places, however, except at specific stations.[2] Each of these stations accommodates one *ATP synthase*, a molecular machine impressively similar to man-made rotary motors, except that, with a velocity of up to several hundred revolutions per second (Ueno et al. 2005) and an efficiency of very nearly 100% (Omote et al. 1999), it outperforms them all. An ATP synthase removed from a bacterium and integrated in a nickel nanopropeller has been shown to power it splendidly (Soong et al. 2000), suggesting that the fabrication of nanomachines from live biological materials is the stuff of science fiction no longer (Strong 2004).

The ATP synthase looks a bit like a mushroom with a rotating stalk embedded within the mitochondrial inner membrane and a head protruding into the matrix. Protons make the stalk rotate by flowing through channels in it, and this movement turns—a third of a full rotation at a time—a stalk-connected rotor sitting within the head. This sequence of turns prompts the head to perform a fixed sequence of operations. At the first turn the head clasps ADP and phosphate; at the second it forces them together to make ATP; at the third it releases the newly formed ATP (Boyer 1975). A full rotation requires ten protons and delivers three molecules of ATP. The ATP can now leave the matrix through a dedicated transporter, embedded in the membrane, that grabs the ATP molecule on the matrix side and deposits it on the other side—a result similar to that achieved by a bank's sliding security doors. The portal is now ready to do the same job in reverse: receive ADP via the open external door, lock it in, and release it on the matrix side by opening the internal door. A separate entrance admits phosphate; upon reaching the matrix, ADP and phosphate are on call to be recharged to ATP again.

When all is well, the electrons flow smoothly down the route to oxygen by hopping from one site to the next, in single file, within each complex. But if they are too slow, the chain becomes clogged up with them; each is unable to move on because the next site is already taken. This can happen, for example, when there is no ADP left in the cell to be converted into ATP, perhaps because ATP is not being consumed fast enough and is thus not broken down into ADP again. An electron stuck in the chain can escape from it onto an oxygen molecule or another chemical nearby. All chemicals contain one or more nuclei, made of protons and neutrons, and a number of electrons orbiting around them. While circling around, each electron spins around its own axis and typically has a companion electron that spins in the opposite direction. By sucking up an extra electron from the electron transport chain, the chemical acquires an unpaired electron, and this turns it into a free radical.

Free radicals are extraordinarily corrosive, first and foremost to the mitochondrion that produces them but also to the cell that hosts the mitochondrion and, by

---

[2]Some protons are allowed to leak back into the matrix through pores in the membrane. This transfer reduces the difference in proton concentration on either side of the membrane and dissipates as heat part of the energy that was pent up in it. In warm-blooded animals, this comes in handy to maintain body temperature (Brand et al. 1991).

extension, to the organism that hosts the cell. Free radicals have a way of reacting with whatever molecules happen to be nearby, including mitochondrial genes which in this way accumulate damage by collecting mutations. An avalanche of free radicals thus bodes no good, but mitochondria have evolved to use it as a smoke alarm—a signal that electrons are not being transferred properly along the chain. In fact, complex I might be deliberately set up so that electrons can leak from it at the slightest hint that the chain is blocked (Lane 2015). In response to a massive release of free radicals within it, a mitochondrion strives to take countermeasures. For example, it can increase the number of respiratory complexes so as to raise ATP production and tailor it to demand, or it can divide and multiply. If demand cannot be met, the inept mitochondrion may go on to commit suicide, freeing space for newcomers; or, if its host cell has incurred too much damage, it may proceed to kill the entire cell. The vital feedback role of free-radical leakage in improving mitochondrial function can help explain, as we shall later see, why antioxidants supplements not only fail to prevent disease but can be downright dangerous.

2 All About Eve

Despite that 99% of mitochondria's components are manufactured by around 1500 nuclear genes and 1% by 13 mitochondrial ones (Rahman and Rahman 2018), it is mitochondrial genes that oversee the control system of energy production (Johnston and Williams 2016). Just like local matters are best left to local rather than national governments, local diversity in energy requirements is best addressed by genes in the locally active mitochondria rather than by those in the faraway nucleus. In response to a smoke alarm, for example, the cell nucleus would have no clue as to which particular mitochondrion is responsible and thus needs to be upgraded or replaced.

Mitochondrial genes do sit very close to where energy is locally produced. Unfortunately, this means they also sit very close to where free radicals are locally released, and hence they might accumulate mutations faster than they would if they were stationed in the nucleus. ("Might," because mitochondrial genes appear to be physically shielded and painstakingly defended from such assaults: Kauppila and Stewart 2015.) Because mitochondria do get heavily tarnished with use, it comes as something of a surprise that those that are inherited by each new generation tend to carry no damage whatsoever. Every bit as surprisingly, the mitochondria that do get passed on—in the vast majority of species including ours—come entirely from mothers (Pyle et al. 2015). Together, human mothers and fathers hand down a set of about 20,000 nuclear genes; in addition, the fertilized egg houses 100,000 live mitochondria (Jansen 2000) exclusively from the mother, each featuring multiple sets of 37 mitochondrial genes.

To make sense of these two surprises, it has been proposed that the mitochondria in egg cells are prevented from ever producing any energy and are simply kept in storage (Allen 1996; Allen and de Paula 2013; de Paula et al. 2013). They will be used as templates for producing new mitochondria after transmission to the off-spring; in female embryos, some of them will be sequestered away in the eggs sitting in the tiny, still developing ovaries. These pristine mitochondria will then be passed on to the next generation—and so on and so forth, from mother to daughter. Eggs may afford to waive active mitochondria because they do nothing much, and can thus subsist on energy coming from less efficient sources or from the mitochondria of adjacent cells in the ovary. Sperm cells, instead, cannot just sit around; they must swim as fast as they possibly can to arrive first and get to fertilize the egg cell. Their mitochondria must deliver a flawless performance, at the cost of getting corroded by free radicals in the process. Thus, whereas egg cells can keep their mitochondria pure and fresh, sperm cells cannot.

To be prevented from being passed on, sperm mitochondria are labeled as unworthy right from the start and are promptly destroyed when they run into the egg's destruction machinery (Sutovsky et al. 1999). Apparently, the harm that they would do to the offspring is so great that fathers cooperate biologically in having them all dismantled upon arrival in the egg. Across species, indeed, a staggering diversity of techniques (Sato and Sato 2013) has evolved to dispense with male mitochondria (including that of defecating them soon after birth: Pitnick and Karr 1998), suggesting that selection pressures favoring uniparental inheritance appeared on the scene repeatedly and independently. One wonders if the problem is only with paternal mitochondria being subpar, or if inheriting two different sets of mitochondrial genes is undesirable in and of itself. Through manipulation and selective breeding it has been possible to answer this question by forcing mice to inherit two perfectly healthy, but different, sets of mitochondrial genes. Interestingly, these mice turn out to be physiologically, emotionally, and cognitively worse off than mice that, other things being equal, inherit just one set (Sharpley et al. 2012).

Assisted reproductive technology may unwittingly force human babies to inherit dissimilar mitochondrial DNA too (e.g., Hiendleder 2007; Sutovsky et al. 1999). Some of these techniques aim to prevent the transmission of mitochondrial disease from mother to child by introducing a twist in the practice of inserting the eggs of a healthy donor into the womb of an unhealthy recipient. Before implantation, the nucleus of each donated egg is replaced with the nucleus removed from an egg of the recipient (Herbert and Turnbull 2018). The recipient can thus pass along to her children all of her own healthy nuclear genes, but none (ideally, because perfection is not of this world) of her unhealthy mitochondrial ones. Babies born with the help of these techniques have effectively three biological parents—having inherited only nuclear DNA from their father, (hopefully) only nuclear DNA from their mother, and only mitochondrial DNA from another, donor mother.

Even disregarding that these babies are likely to get some of their mitochondria from the wrong mother, the trouble is that the nuclear genome from one mother and the mitochondrial genome from the other might not cooperate properly (Hiendleder 2007). The apparently risible percentage of mitochondrial components encoded by the mitochondrial genome includes most of the building blocks of the electron transport chain; the remaining are produced by the nuclear genome (Johnston and Williams 2016). A less than perfect interaction between the two genomes might alter the nanoscopically precise distances between the sites through which the electrons pass down the route to oxygen, impairing their flow—with more or less catastrophic consequences (Lane 2011). Only time will tell how the health and lifespan of people with three parents measure up to those with just two.

Incidentally, the fact that mitochondrial (unlike nuclear) DNA never gets mixed with that of fathers makes it uniquely suitable for tracing people's ancestry along the female line. Following our progenitors' mitochondrial footprints back from mother to grandmother to great-grandmother, one arrives at Mitochondrial Eve, a title currently held by a woman who gave birth to her own daughters around 200,000 years ago (Cann et al. 1987) and from whom all people alive today descend in an unbroken female line.[3]

3 All Creatures Great and Small

How much an organism needs to eat depends on how fast it converts food into energy. This metabolic rate depends on whether the organism is moving or not and on what its characteristics are, such as whether it is cold or warm blooded, what sexual strategy it likes better, and how big it is (Lane 2005). Yet not all body parts demand an amount of energy that is proportional to the organism's overall size (Haldane 1926). In larger land mammals, for example, bones—which require little energy to maintain—are relatively wider, but the blood vessel system is less dense, with each capillary feeding a greater number of cells, and this allows the heart to beat more slowly, saving energy (Lane 2005). Indeed, larger species spend energy more efficiently: food and oxygen consumption per gram of body weight diminish with size, along with the breathing, heartbeat, and metabolic rates (Lane 2005). The lower the metabolic rate, the lower the production of free radicals per cell and per mitochondrion—and thus the lower the risk of early degeneration and death. Indeed, large species tend to outlive small ones (Blagosklonny 2013): a lucky mouse has a life expectancy of 2 or 3 years, an elephant of 60 or 70.

Large species spend their energy more efficiently than small ones, grow to maturity more slowly, and—perhaps counterintuitively—feature lower levels of growth hormones (Stuart and Page 2010; Swanson and Dantzer 2014; Tian et al. 2017). Within a species, however, all individuals mature about equally fast, thus larger adults must have grown faster than smaller ones. Individuals' final size is likely to reflect an inclination towards promoting or inhibiting the hormonal mechanisms that sustain growth; indeed, larger adults have higher levels of growth hormones (Bartke 2017; Swanson and Dantzer 2014).

The problem with growth is that it depends on cell divisions and each division requires the copying of DNA. Every round of copying increases the probability of mistakes (mutations) and, with it, the probability that growth spins out of control, leading to cancer. Larger species, that in principle would be most vulnerable to

[3]Other women were around at the time and also earlier, of course, but their own mitochondrial lineages happened to become extinct. Some of these women may well have descendants alive today, yet not in a direct female line but passing, at some point or other, through a male (a dead end for mitochondria).

cancer, actually invest more than do smaller species in mechanisms dedicated to fighting it (Tian et al. 2017). For example, elephants have 20 copies of a gene that forces cancer cells to commit suicide, whereas humans and mice have 1 copy only (Sulak et al. 2016). As it happens, genetically modified mice with an extra copy develop fewer tumors (Matheu et al. 2007).

Whereas individuals of larger species are better protected against cancer than those of smaller ones—which cancels the costs of being big—larger and smaller individuals of the *same* species have identical defenses in place. In line with this, taller men are more likely to suffer from cancer, and tend in general to die earlier, than shorter men (He et al. 2014). Tellingly, the manufacturing of an important growth hormone appears downregulated in both human centenarians (Suh et al. 2008) and their offspring (Vitale et al. 2012). If growth hormones are diminished artificially, an experiment that has been tried in a kind of roundworm (Zarse et al. 2012), cells suck in less glucose from the blood; starved of sugar, they must switch to less preferred fuels, such as those derived from fat. So, they have no choice but to rely even more heavily on mitochondria than they usually do, because only mitochondria can make ATP by burning fat. This surge in mitochondrial activity is accompanied by a modest increment in free radical levels, which in turn prompts antioxidant defenses that reduce these levels to values below those seen before the intervention. The net effect is that the worms age more slowly and end up living twice as long as usual (see also Miller et al. 2018; Schulz et al. 2007).

One can achieve similar results in creatures other than worms by simply offering them no more food than necessary. In rodents, for example, restricting food intake reduces the level of the chemical that feeds electrons into the electron transport chain, the escape of electrons from the chain, and the formation of free radicals (Barja 2013). As a rule, eating adequately but less than desired postpones the diseases of aging and prolongs life in organisms ranging from yeast to fish, from spiders to humans (Fontana et al. 2010). In humans, these benefits also apply to people who are not considered overweight to start with (Redman et al. 2018).

Eating less than preferred is very hard. With this truth in mind, dietary regimes have been proposed that do not restrict calories but still lower the levels of glucose, like ketogenic diets, or those of growth hormones, like diets that curtail the consumption of protein or some critical component of it. Ketogenic diets feature plenty of fats such as cream and butter but very little carbohydrates, our main source of glucose, and thus conveniently mimic fasting (Longo and Mattson 2014). Lack of sugar pushes the liver to break down stored fat into molecules called ketone bodies, which mitochondria can use for fuel (e.g., Newman and Verdin 2014). Ketone bodies, likely as an adaptive response to nutrient scarcity, also intensify the recycling of damaged cell components and similar debris, which may serve as emergency food for the cell (Camberos-Luna et al. 2016). Perhaps because it removes potentially lethal garbage, this frenetic housekeeping might be another reason why ketogenic diets appear to ameliorate a variety of afflictions, ranging from neurological diseases to cancer (Branco et al. 2016; Gano et al. 2014; Kramer and Bressan 2018; Miller et al. 2018).

Ketogenic diets drastically cut out carbohydrates but reduce protein too, on the grounds that protein could otherwise be converted into glucose. This measure may have an accidental further advantage, because restricting protein—or at least two of its most common component amino acids—appears to promote health and longevity in itself (Dong et al. 2018). These two amino acids are methionine and cysteine (methionine restriction being largely ineffective if cysteine is not removed too: Wanders et al. 2015). Methionine is an essential amino acid, one that cannot be produced by the body but must be introduced with the diet, and the ability to sense its deficiency and promptly respond to it is of course indispensable to survival. As is often the case with indispensable activities, multiple mechanisms appear to be in place to achieve the same purpose. A major one involves the liver, which within hours of a methionine-restricted meal increases fivefold its production of a fundamental metabolic regulator (Forney et al. 2017a).

This regulator proceeds to orchestrate a suite of adaptive responses that include a coordinated increase in energy intake (likely on the lookout for extra methionine, one eats more) and expenditure (Anthony et al. 2013). The latter is achieved mainly by making the mitochondrial inner membrane leakier to protons, so that some of the energy pent up in the intermembrane space, rather than being used to make ATP, is wasted as heat (Hasek et al. 2010). As a consequence, the production of free radicals drops, and with it drops the damage that free radicals inflict on tissues and DNA (e.g., López-Torres and Barja 2008). It may be no coincidence that species that live longer, such as the pigeon, happen to carry less methionine in their flesh than species that live shorter, such as the rat. In mammals—from mice to horse, through rabbit and pig—the negative correlation between methionine content in the heart and maximum lifespan is nearly perfect (Pamplona and Barja 2006).

Despite ingesting more food, animals on methionine-restricted diets stay thin because the cost of growth increases: more energy is required for maintaining existing tissue (Hasek et al. 2010). Yet a cost is a cost, and while the degree of restriction that compels individuals to take in more food proves salutary, going just a bit further down that road causes excessive weight loss: animals become thinner and thinner but stop consuming extra food, as though methionine is now so low that eating more would make no difference (Forney et al. 2017b). An even stricter dilution of any essential amino acid creates (within minutes: Hao et al. 2005) an aversion to the unbalanced food, and, if no alternative source can be tracked down, simply fails to sustain life. More of a good thing is not always a better thing.

4 Have Kids, Die Early

Periodic famines may be one reason why we animals eat more than we need. Another may be that stern diets inhibit the immune system, impairing wound healing and recovery from infection; reduce the response to cold, increasing mortality in small animals when temperature drops; and diminish bone and muscle mass, potentially hindering one's ability to fight or escape from predators (Adler and

Bonduriansky 2014). All of this saves energy, but—except under lab conditions—bodes ill for survival. Yet the most important reason why we desire more food than necessary may be that eating is not just about sustaining ourselves but also about producing offspring. The hormones and enzymes that encourage growth in response to nutrient abundance reduce lifespan but also upregulate hormonal mechanisms that bolster fertility and sexual activity (follicle-stimulating hormone, luteinizing hormone, estrogen, and testosterone); hormones and enzymes that reduce growth do the opposite (Ellison 2017; Templeman and Murphy 2018).

Within species, individuals that invest more in acquiring mates and producing offspring tend to die earlier (Lemaître et al. 2015). This is also the case in humans: reproductive success (measured as either a larger number of children or shorter birth intervals) is associated with reduced survival in both women and, to a smaller extent, men—even after adjusting for socioeconomic status and other potential confounds (Penn and Smith 2007). Rounding off this evidence, both male and female centenarians have fewer children than other people and had them later in life (shown in Ashkenazi Jews: Tabatabaie et al. 2011). Thus, the survival costs of reproduction go beyond the physical burdens of pregnancy, childbirth, and breastfeeding.

Achieving reproductive maturity requires development and growth, and the amount of growth that has taken place is reflected in individuals' size. Indeed, dwarf people resemble centenarians (who, as mentioned earlier, manufacture less growth hormone) in that they reproduce later in life (Bartke et al. 2013; Hull and Harvey 2014). In dwarfs, the odds of developing diabetes and cancer are also lower than average, whereas in giants they are higher and accompanied by a reduced life expectancy (Bartke 2017). In mice bred to be dwarfs, which do not have enough of a particular growth hormone, fertility is either abolished or delayed, but lifespan is stretched by up to 50% (Bartke et al. 2013). Female hyenas with low levels of growth hormone in their youth tend to be smaller as adults, but even when they are not, they breed later and live longer (Lewin et al. 2017).

Intriguingly, in humans (like in several other animals), low maternal food intake has been associated with a greater likelihood of giving birth to females than to males, and women who gain less weight during gestation produce fewer sons (Navara 2014; see also Myers 1978; Trivers and Willard 1973). The main reason appears to be that sons tend to be larger than daughters and hence require more resources, even in the womb: more male than female fetuses are spontaneously aborted when food is scarce. Women who became pregnant during the great Chinese famine from 1957 to 1961, which killed 30 million people, were less likely to give birth to sons than expected (Song 2012, 2015); the same happened during the 1974 famine in Bangladesh (Hernández-Julián et al. 2014).

Growth, reproduction, and longevity are also interrelated in another interesting way. Different species have different strategies regarding the quality and quantity of their offspring (Lane 2015). Those enforcing the strictest quality control discard as defective the most egg cells and spontaneously abort the most of their potential offspring. This results in fewer but healthier offspring that can survive longer. Those producing more offspring are forced to lower their quality control standards, but then their offspring tend to die sooner. It is a species' circumstances that appear to dictate its strategy. For example, creatures that fly require more energy than those that do not, and this calls for better mitochondria and therefore for higher quality control of offspring. A pigeon delivers 15 young per year and a rat 60, but then the pigeon's lifespan is ten times longer than the rat's.

Demonstrating that mitochondria play a critical part in such survival strategies, beetles that are artificially selected for either long or short lives, and respectively breed later or earlier, end up with distinctly different types of mitochondrial DNA (Stojković et al. 2017). From an evolutionary point of view, of course, it is better to live a short life yet reproduce than to live a long one but leave no descendants. In this respect, the fact that mitochondrial DNA does not care about its effects on males has occasionally a silver lining for them. For example, a type of mitochondrial DNA found in fruit flies renders females fertile but short-lived and males sterile but long-lasting (Camus et al. 2015).

5 Guarding the Gate

We are not alone in our bodies; we house some 38 trillion microbes (Sender et al. 2016) belonging to at least 2172 known species (Hugon et al. 2015). Our health depends also on theirs (Gilbert et al. 2016; Kramer and Bressan 2015; Rogers et al. 2016) and so does the health of our mitochondria. Most of these microbes live in our gut and are heavily affected by the food we eat. For example, protein is mostly digested by enzymes in our nearly sterile small intestine. The rest progresses to the colon, where it is fermented by bacteria that expel, as waste products, toxins like hydrogen sulfide (Linden 2014; Pal et al. 2018; Yao et al. 2016)—the gas that gives rotten eggs their foul smell. Some vegetables, like cauliflower and broccoli, have evolved an abundance of sulfur compounds to ward off their own parasites (Björkman et al. 2011). Such compounds harm the bacteria in our gut that disgorge hydrogen sulfide (Kellingray et al. 2017), opening up opportunities for rival, beneficial ones; which might be part of the reason why a regular consumption of this family of vegetables is associated with a lower probability of developing diabetes, cardiovascular diseases, and cancer (Björkman et al. 2011).

Hydrogen sulfide is not entirely bad; some of our cells assemble it in small amounts from cysteine and use it, for example, to send important messages to other cells (Banerjee et al. 2015). Still, the more protein we eat, the more hydrogen sulfide we get; eventually, it may become too much. If so, the gas can tear open the mucus layer that protects the gut wall, breach into the wall's cells, and harm their mitochondria (Blachier et al. 2007; Mottawea et al. 2016). This chain of events raises the mitochondria's free radical production—which in turn increases the permeability of the barrier that separates the gut from the bloodstream (Mottawea et al. 2016), granting entry into the body to those microbes, viruses (Pal et al. 2018), and potentially immunoreactive food particles (Bressan and Kramer 2016) that are normally held back inside the gut. It is no wonder that specific assortments of gut microbes have been associated with disorders of every description (Bastiaanssen et al. 2018; Gilbert et al. 2016; Kramer and Bressan 2015; Rogers et al. 2016).

The task of keeping pathogens confined within the gut is delegated to the immune cells that line its inside. Curiously, these cells get their energy from mitochondria whose electron transport chains are embedded in an unusual variant of the inner membrane—one that is purposely fashioned to *limit* energy production (Konjar et al. 2018). In response to the inflammatory signals triggered by an infection, the properties of this membrane suddenly change: the brakes come off and energy production shoots up to full capacity. Now the immune cells get a jolt of energy and multiply rapidly to attack any pathogens that may be hanging about. The concomitant increase in free radical levels, which is indeed typical of inflammation (Angajala et al. 2018; López-Armada et al. 2013), helps wrecking the pathogens further. The mitochondria with the unusual membranes are thus effectively gatekeepers: they keep a large number of guard dogs (the immune cells) under control when all is well, but set them loose when evil characters (the pathogens) arrive on the scene. With the gut's immune cells under the control of mitochondria and the immune cells themselves always at beck and call to assail gut microbes, it makes a lot of sense that differences in mitochondrial DNA are associated with differences in gut microbe composition (Clark and Mach 2017; Hirose et al. 2017). In an overt attempt to overcome an immune response against them—a response which is invariably assisted by a hefty release of free radicals—several pathogens discharge toxins that interfere with their host's mitochondrial signaling and overactivate its antioxidant mechanisms (Clark and Mach 2017).

Indeed, every so often our frontline troops are not up to the task and we need to destroy the pathogens ourselves. In most societies, few go through life without ever taking antibiotics: globally, just over the period 2000–2015, their consumption has increased by 65%, reaching nearly 35 billion doses per day (Klein et al. 2018). Although they can be indispensable occasionally, antibiotics famously wipe out beneficial gut bacteria along with offensive ones—allies and enemies both—and worsen the balance between them after recovery (Keeney et al. 2014). On top of that, antibiotics harm mitochondria very directly, because of course mitochondria resemble bacteria, their forebears. Even at recommended doses for the recommended time, antibiotics injure the outer wall and DNA of mitochondria, upset their quality control activities, and disrupt electron transport chains—making them produce less energy and more free radicals (Kalghatgi et al. 2013; Wang et al. 2015). Whatever hurts microbes—most prominently, common pesticides and pollutants (Karami-Mohajeri and Abdollahi 2013; Meyer et al. 2013; Mostafalou and Abdollahi 2013)—is indeed likely to hurt mitochondria. In our fight against pathogens and pests, we ought to be much more careful about collateral damage to our tenants.

6 A Call to (Some) Action

During the day we eat; digested food is stripped of electrons and these are fed into the transport chain to make fresh ATP. Yet if we do not move, not much ATP is broken down into ADP, and our ATP synthases run out of supplies. When ADP is unavailable, the ATP synthase stops and protons can no longer flow, through it, into the matrix (Wallace 2005). As the protons accumulate inside the intermembrane space, their increasing concentration impedes further pumping of protons into this space. The pumping slows down, and the stream of electrons through the electron transport chain slows down with it. Electrons continue to be fed into the chain, however; as they heap up, more and more of them escape before reaching their final destination (shown in vitro: Herrero and Barja 1997), react with oxygen, and form free radicals.

Thus, by consuming ATP and hence speeding up the flow of electrons, exercise decreases (rather than increases, as one might have thought), the formation of free radicals *inside* mitochondria. But mitochondria are not the only source of free radicals; these are fabricated in various chemical reactions elsewhere too, and during exercise their net level goes up, not down (He et al. 2016; Powers and Jackson 2008; Trewin et al. 2018). If one exercises regularly, the body puts into effect a suite of countermeasures to be ready to cope with similar stresses in the future (Goto et al. 2007; Gradari et al. 2016; Hood et al. 2011). Such countermeasures include the upregulation of repair mechanisms and the production of both antioxidants and enzymes meant to get more glucose out of the blood and into cells (reducing, incidentally, the risk of diabetes). They also include an upgrade in the number of mitochondria and of the electron transport chains inside them. This has an antioxidant effect because the more chains there are, the fewer electrons can be fed into each—per unit of food burned down—and thus the fewer electrons leak out to form free radicals. Somewhat counterintuitively, then, the beneficial effects of exercise turn out to be due (albeit indirectly) to a *rise* in free radical levels.

Motivated by the wish to do the right thing, many healthy people—athletes prominently included—exercise (which raises free radical levels) and take antioxidant supplements (which lowers them). Yet antioxidant supplements turn out to *block* the health-promoting effect of exercise (Merry and Ristow 2016; Peternelj and Coombes 2011; Ristow et al. 2009; Villanueva and Kross 2012). The harmful effects of the free radicals produced by exercise are compensated—surpassed, in fact—by the countermeasures they trigger, and this is why exercise is healthy after all.

Note that this holds when exercise is moderate. When it is excessive, free radical levels may increase too much for benefits to outweigh costs (in which case taking antioxidants might even be fine: Radak et al. 2017, though not wearing oneself out seems like a better plan). As it happens, rats that are forced to swim either 10 or 30 min a day for 20 weeks develop fewer mutations in their mitochondrial DNA than rats who remain idle (Cao et al. 2012). Yet, whereas rats that swim for 10 or, better, 30 min a day end up with more copies of mitochondrial genes (associated with greater resilience), rats that swim for 60 or 90 min end up with fewer. Similarly, in humans, overtraining has been associated with considerable free-radical damage to fat tissues, proteins, and DNA (Radak et al. 2008, 2013).

It stands to reason, then, that the effects of exercise depend on how much one is biologically capable of generating energy and on how easily one produces free radicals. Both traits are determined, in part, by the type of mitochondrial DNA one has inherited from one's mother. For example, carriers of one type manufacture more

ATP and free radicals than carriers of another (Martínez-Redondo et al. 2010). Interestingly, Olympic endurance athletes (such as swimmers, cyclists, long-distance runners, and cross-country skiers: Maruszak et al. 2014) tend to carry the former type.

7 There Is a Time for Everything

Unexpectedly, the production of energy and regulation of free radical levels have popped up as crucial in the supervision of a still largely mysterious activity in which we engage about one-third of our time: sleep. For the past decades, researchers of biorhythms (recurring cycles in biological activity) have been focusing their attention mostly on a central biological clock sitting in the hypothalamus and on the genes in charge of it (Borbély et al. 2016; Rutter et al. 2002). To keep track of time, the clock accumulates certain chemicals until a threshold is reached and then it starts afresh. Presumably, this central clock allows organisms to prepare for, rather than merely react to, recurring changes in light and temperature, so that they can anticipate opportunities to eat and engage in fruitful activities. To this effect, it adjusts the release of hormones that are responsible for one's inclination to eat during the day and sleep during the night (Laje et al. 2018). Although it induces a sleep/wake cycle regardless of the external, night/day one, the clock manages to keep the two cycles in sync by using a report, issued by our eyes, of the changes in overall illumination.

If people are given full control over artificial lighting but are deprived of natural light, their patterns of eating, moving, and sleeping change—sometimes drastically, with wake periods shorter than 12 h or longer than 30 (Aschoff et al. 1986). Remarkably, these alterations mirror changes in energy intake and expenditure (Borbély et al. 2016; Reddy and Rey 2014). People who spontaneously slow down their day from 24 to 48 h, for example, proportionally slow down their food intake from three meals per 24-h day to three per 48-h day (Aschoff et al. 1986; Green et al. 1987). Their total amount of activity slows down proportionally too (Aschoff 1993). In these conditions, "warmer" people (those whose mean body temperature is higher) tend to shorten their sleep/wake cycle whilst "cooler" people tend to lengthen it (Daan et al. 2013). A tiny variation in core temperature (0.5 °C) corresponds to hugely different individual day lengths—spanning from 16 to 48 h. This agrees with the notion that the alternation of sleep and wake strongly depends on energy management.

The central clock regulates other clocks that are stationed throughout the body—in the liver, in the heart and such. Both the central clock and these local ones speed up when the level of an important nutrient-sensing (and growth-promoting) enzyme rises, and slow down when it falls (shown in mice: Ramanathan et al. 2018). Interestingly, exposure to light can reset one's central clock via the same nutrient-sensing enzyme; inhibiting the enzyme blocks this reset (Cao et al. 2010; see also Cao et al. 2008).

Indeed, energy management may be more important for keeping track of time than any central or local clocks. In red blood cells kept in vitro in complete darkness, for example, the activity of antioxidant mechanisms and the level of ATP turn out to wax and wane with a 24-h circadian rhythm (O'Neill and Reddy 2011). This is an impressive finding, because red blood cells lack a nucleus and with it the genes necessary to maintain a local clock; sitting in a Petri dish, they cannot possibly be affected by a central brain clock either—which suggests that biorhythms are driven by energy production *per se*.

Red blood cells lack not only a nucleus but mitochondria too; they rely on an alternative, less efficient source of energy. In most of our cells, however, it is primarily the mitochondria that produce energy. One may thus suppose that it is primarily the mitochondria that control our biorhythms, including the sleep/wake cycle. Consistent with this idea, rats whose brain is infused with low levels of a free-radical-producing chemical fall asleep more often and sleep longer (Ikeda et al. 2005).

The burning of digested food in mitochondria increases and decreases with recurrent ups and downs in food intake and therefore with the alternation of wake and sleep. The cyclic production of energy is accompanied by a cyclic production of free radicals and a cyclic need to neutralize them, and it appears that sleep is part of a larger mechanism to accomplish this feat (Ikeda et al. 2005). As a matter of fact, glutathione—an important antioxidant in many organisms—induces sleep (Inoué et al. 1995) and melatonin—a well-known inducer of sleep—is not only a hormone but also an antioxidant. Melatonin is plentiful in mitochondria, and it has been argued that the bacterial ancestors of mitochondria might have produced melatonin as an antioxidant before it ever became a sleep hormone in creatures like us (Reiter et al. 2017).

The importance of our biorhythms is well illustrated by what happens when we mess with them. The invention of artificial light has allowed people to be active any time of day or night. Yet, whether it is due to going to bed too late, to jet lag, or most prominently to shift work, being awake or asleep at the wrong time has been associated with a myriad physical and mental disorders, including anxiety, depression, obesity, heart disease, and cancer (Kondratova and Kondratov 2012; Liu and Chung 2015; Roenneberg and Merrow 2016). Reversing the sleep/wake cycle of human volunteers for just 2 days disrupts the normal levels and daily variation patterns of more than 100 proteins in the blood, including several that are involved in the regulation of blood sugar, weight gain, immune function, and cancer (Depner et al. 2018).

8 Coda: Have Your Cake or Eat It

Mitochondria produce nearly all of the energy we animals need. We have evolved to invest this energy to grow, find food and shelter, and leave descendants, and to last long enough to make it all happen. Yet because the continuous manufacturing of energy is inevitably linked to wear-and-tear, mitochondria give us death as well as life. And because energy spent on growing and procreating is energy not spent on recycling and repairing, producing many offspring and living a long life trade off with each other.

What compromises are made in our own special case (how big we become, how early we can reproduce, whether we happen to be born women or men, or for that matter pigeons or rats) is very largely beyond our control. So is the unappetizing truth that hordes of uninvited guests (microbes, viruses, and other selfish entities) are feasting on our resources too, shaping all along how we function and behave (Kramer and Bressan 2015). What we *can* do is treat our mitochondria as allies and adjust our lifestyle so as to take good care of their necessities. By eating properly, exercising regularly, and sleeping at night, we can keep our companions' proton pumps well oiled and their ATP synthases plugging away. We may get in exchange the energy we need to live interestingly.

References

Adler MI, Bonduriansky R (2014) Why do the well-fed appear to die young? Bioessays 36:439–450. https://doi.org/10.1002/bies.201300165

Allen JF (1996) Separate sexes and the mitochondrial theory of ageing. J Theor Biol 180:135–140. https://doi.org/10.1006/jtbi.1996.0089

Allen JF, de Paula WBM (2013) Mitochondrial genome function and maternal inheritance. Biochem Soc Trans 41:1298–1304. https://doi.org/10.1042/BST20130106

Angajala A, Lim S, Phillips JB et al (2018) Diverse roles of mitochondria in immune responses: novel insights into immuno-metabolism. Front Immunol 9:1605. https://doi.org/10.3389/fimmu.2018.01605

Anthony TG, Morrison CD, Gettys TW (2013) Remodeling of lipid metabolism by dietary restriction of essential amino acids. Diabetes 62:2635–2644. https://doi.org/10.2337/db12-1613

Aschoff J (1993) On the relationship between motor activity and the sleep-wake cycle in humans during temporal isolation. J Biol Rhythm 8:33–46. https://doi.org/10.1177/074873049300800103

Aschoff J, von Goetz C, Wildgruber C et al (1986) Meal timing in humans during isolation without time cues. J Biol Rhythm 1:151–162. https://doi.org/10.1177/074873048600100206

Banerjee R, Chiku T, Kabil O et al (2015) Assay methods for H2S biogenesis and catabolism enzymes. In: Cadenas E, Packer L (eds) Hydrogen sulfide in redox biology, part A. Methods in enzymology, vol 554. Elsevier, Amsterdam, pp 189–200

Barja G (2013) Updating the mitochondrial free radical theory of aging: an integrated view, key aspects, and confounding concepts. Antioxid Redox Signal 19:1420–1445. https://doi.org/10.1089/ars.2012.5148

Bartke A (2017) Somatic growth, aging, and longevity. NPJ Aging Mech Dis 3:14. https://doi.org/10.1038/s41514-017-0014-y

Bartke A, Sun LY, Longo V (2013) Somatotropic signaling: trade-offs between growth, reproductive development, and longevity. Physiol Rev 93:571–598. https://doi.org/10.1152/physrev.00006.2012

Bastiaanssen TF, Cowan CS, Claesson MJ et al (2018) Making sense of … the microbiome in psychiatry. Int J Neuropsychopharmacol. https://doi.org/10.1093/ijnp/pyy067/25437947/pyy067.pdf

Björkman M, Klingen I, Birch AN et al (2011) Phytochemicals of Brassicaceae in plant protection and human health – influences of climate, environment and agronomic practice. Phytochemistry 72:538–556. https://doi.org/10.1016/j.phytochem.2011.01.014
Blachier F, Mariotti F, Huneau JF et al (2007) Effects of amino acid-derived luminal metabolites on the colonic epithelium and physiopathological consequences. Amino Acids 33:547–562. https://doi.org/10.1007/s00726-006-0477-9
Blagosklonny MV (2013) Big mice die young but large animals live longer. Aging 5:227–233. https://doi.org/10.18632/aging.100551
Borbély AA, Daan S, Wirz-Justice A et al (2016) The two-process model of sleep regulation: a reappraisal. J Sleep Res 25:131–143. https://doi.org/10.1111/jsr.12371
Boyer PD (1975) A model for conformational coupling of membrane potential and proton translocation to ATP synthesis and to active transport. FEBS Lett 58:1–6
Branco AF, Ferreira A, Simões RF et al (2016) Ketogenic diets: from cancer to mitochondrial diseases and beyond. Eur J Clin Investig 46:285–298. https://doi.org/10.1111/eci.12591
Brand MD, Couture P, Else PL et al (1991) Evolution of energy metabolism. Proton permeability of the inner membrane of liver mitochondria is greater in a mammal than in a reptile. Biochem J 275:81–86
Bressan P (2018) Inside a mitochondrion [Figure]. Figshare. https://doi.org/10.6084/m9.figshare.6877205. https://figshare.com/s/5d90908b41fe3524efc0. Accessed 1 Aug 2018
Bressan P, Kramer P (2016) Bread and other edible agents of mental disease. Front Hum Neurosci 10:130
Camberos-Luna L, Gerónimo-Olvera C, Montiel T et al (2016) The ketone body, β-hydroxybutyrate stimulates the autophagic flux and prevents neuronal death induced by glucose deprivation in cortical cultured neurons. Neurochem Res 41:600–609. https://doi.org/10.1007/s11064-015-1700-4
Camus MF, Wolf JB, Morrow EH et al (2015) Single nucleotides in the mtDNA sequence modify mitochondrial molecular function and are associated with sex-specific effects on fertility and aging. Curr Biol 25:2717–2722. https://doi.org/10.1016/j.cub.2015.09.012
Cann RL, Stoneking M, Wilson AC (1987) Mitochondrial DNA and human evolution. Nature 325:31–36. https://doi.org/10.1038/325031a0
Cao R, Lee B, Cho H-Y et al (2008) Photic regulation of the mTOR signaling pathway in the suprachiasmatic circadian clock. Mol Cell Neurosci 38:312–324. https://doi.org/10.1016/j.mcn.2008.03.005
Cao R, Li A, Cho H-Y et al (2010) Mammalian target of rapamycin signaling modulates photic entrainment of the suprachiasmatic circadian clock. J Neurosci 30:6302–6314. https://doi.org/10.1523/jneurosci.5482-09.2010
Cao X, Zhao Z-W, Zhou H-Y et al (2012) Effects of exercise intensity on copy number and mutations of mitochondrial DNA in gastrocnemus muscles in mice. Mol Med Rep 6:426–428. https://doi.org/10.3892/mmr.2012.913
Clark A, Mach N (2017) The crosstalk between the gut microbiota and mitochondria during exercise. Front Physiol 8:319. https://doi.org/10.3389/fphys.2017.00319
Daan S, Honma S, Honma K-I (2013) Body temperature predicts the direction of internal desynchronization in humans isolated from time cues. J Biol Rhythm 28:403–411. https://doi.org/10.1177/0748730413514357
de Paula WBM, Lucas CH, Agip A-NA et al (2013) Energy, ageing, fidelity and sex: oocyte mitochondrial DNA as a protected genetic template. Philos Trans R Soc Lond Ser B Biol Sci 368:20120263. https://doi.org/10.1098/rstb.2012.0263
Depner CM, Melanson EL, McHill AW et al (2018) Mistimed food intake and sleep alters 24-hour time-of-day patterns of the human plasma proteome. Proc Natl Acad Sci U S A 115:E5390–E5399. https://doi.org/10.1073/pnas.1714813115
Dong Z, Sinha R, Richie JP (2018) Disease prevention and delayed aging by dietary sulfur amino acid restriction: translational implications. Ann N Y Acad Sci 1418:44–55. https://doi.org/10.1111/nyas.13584

Ellison PT (2017) Endocrinology, energetics, and human life history: a synthetic model. Horm Behav 91:97–106. https://doi.org/10.1016/j.yhbeh.2016.09.006
Fontana L, Partridge L, Longo VD (2010) Extending healthy life span – from yeast to humans. Science 328:321–326. https://doi.org/10.1126/science.1172539
Forney LA, Stone KP, Wanders D et al (2017a) Sensing and signaling mechanisms linking dietary methionine restriction to the behavioral and physiological components of the response. Front Neuroendocrinol 51:36–45. https://doi.org/10.1016/j.yfrne.2017.12.002
Forney LA, Wanders D, Stone KP et al (2017b) Concentration-dependent linkage of dietary methionine restriction to the components of its metabolic phenotype. Obesity 25:730–738. https://doi.org/10.1002/oby.21806
Gano LB, Patel M, Rho JM (2014) Ketogenic diets, mitochondria, and neurological diseases. J Lipid Res 55:2211–2228
Gilbert JA, Quinn RA, Debelius J et al (2016) Microbiome-wide association studies link dynamic microbial consortia to disease. Nature 535:94–103. https://doi.org/10.1038/nature18850
Goto S, Naito H, Kaneko T et al (2007) Hormetic effects of regular exercise in aging: correlation with oxidative stress. Appl Physiol Nutr Metab 32:948–953. https://doi.org/10.1139/H07-092
Gradari S, Pallé A, McGreevy KR et al (2016) Can exercise make you smarter, happier, and have more neurons? A hormetic perspective. Front Neurosci 10:93. https://doi.org/10.3389/fnins.2016.00093
Green J, Pollak CP, Smith GP (1987) Meal size and intermeal interval in human subjects in time isolation. Physiol Behav 41:141–147
Haldane JBS (1926) On being the right size. Harper Mag 152:424–427
Hao S, Sharp JW, Ross-Inta CM et al (2005) Uncharged tRNA and sensing of amino acid deficiency in mammalian piriform cortex. Science 307:1776–1778. https://doi.org/10.1126/science.1104882
Hasek BE, Stewart LK, Henagan TM et al (2010) Dietary methionine restriction enhances metabolic flexibility and increases uncoupled respiration in both fed and fasted states. Am J Physiol Regul Integr Comp Physiol 299:R728–R739. https://doi.org/10.1152/ajpregu.00837.2009
He Q, Morris BJ, Grove JS et al (2014) Shorter men live longer: association of height with longevity and FOXO3 genotype in American men of Japanese ancestry. PLoS One 9:e94385. https://doi.org/10.1371/journal.pone.0094385
He F, Li J, Liu Z et al (2016) Redox mechanism of reactive oxygen species in exercise. Front Physiol 7:486. https://doi.org/10.3389/fphys.2016.00486
Herbert M, Turnbull D (2018) Progress in mitochondrial replacement therapies. Nat Rev Mol Cell Biol 19:71–72. https://doi.org/10.1038/nrm.2018.3
Hernández-Julián R, Mansour H, Peters C (2014) The effects of intrauterine malnutrition on birth and fertility outcomes: evidence from the 1974 Bangladesh famine. Demography 51:1775–1796. https://doi.org/10.1007/s13524-014-0326-5
Herrero A, Barja G (1997) ADP-regulation of mitochondrial free radical production is different with complex I- or complex II-linked substrates: implications for the exercise paradox and brain hypermetabolism. J Bioenerg Biomembr 29:241–249
Hiendleder S (2007) Mitochondrial DNA inheritance after SCNT. Adv Exp Med Biol 591:103–116. https://doi.org/10.1007/978-0-387-37754-4_8
Hirose M, Künstner A, Schilf P et al (2017) Mitochondrial gene polymorphism is associated with gut microbial communities in mice. Sci Rep 7:15293. https://doi.org/10.1038/s41598-017-15377-7
Hood DA, Uguccioni G, Vainshtein A et al (2011) Mechanisms of exercise-induced mitochondrial biogenesis in skeletal muscle: implications for health and disease. Compr Physiol 1:1119–1134. https://doi.org/10.1002/cphy.c100074
Hugon P, Dufour J-C, Colson P et al (2015) A comprehensive repertoire of prokaryotic species identified in human beings. Lancet Infect Dis 15:1211–1219. https://doi.org/10.1016/s1473-3099(15)00293-5
Hull KL, Harvey S (2014) Growth hormone and reproduction: a review of endocrine and autocrine/paracrine interactions. Int J Endocrinol 2014:1–24. https://doi.org/10.1155/2014/234014

Ikeda M, Ikeda-Sagara M, Okada T et al (2005) Brain oxidation is an initial process in sleep induction. Neuroscience 130:1029–1040. https://doi.org/10.1016/j.neuroscience.2004.09.057
Inoué S, Honda K, Komoda Y (1995) Sleep as neuronal detoxification and restitution. Behav Brain Res 69:91–96
Jansen RP (2000) Germline passage of mitochondria: quantitative considerations and possible embryological sequelae. Hum Reprod 15(Suppl 2):112–128
Johnston IG, Williams BP (2016) Evolutionary inference across eukaryotes identifies specific pressures favoring mitochondrial gene retention. Cell Syst 2:101–111. https://doi.org/10.1016/j.cels.2016.01.013
Kalghatgi S, Spina CS, Costello JC et al (2013) Bactericidal antibiotics induce mitochondrial dysfunction and oxidative damage in mammalian cells. Sci Transl Med 5:192ra85. https://doi.org/10.1126/scitranslmed.3006055
Karami-Mohajeri S, Abdollahi M (2013) Mitochondrial dysfunction and organophosphorus compounds. Toxicol Appl Pharmacol 270:39–44. https://doi.org/10.1016/j.taap.2013.04.001
Kauppila JH, Stewart JB (2015) Mitochondrial DNA: radically free of free-radical driven mutations. Biochim Biophys Acta 1847:1354–1361. https://doi.org/10.1016/j.bbabio.2015.06.001
Keeney KM, Yurist-Doutsch S, Arrieta MC et al (2014) Effects of antibiotics on human microbiota and subsequent disease. Annu Rev Microbiol 68:217–235. https://doi.org/10.1146/annurev-micro-091313-103456
Kellingray L, Tapp HS, Saha S et al (2017) Consumption of a diet rich in Brassica vegetables is associated with a reduced abundance of sulphate-reducing bacteria: a randomised crossover study. Mol Nutr Food Res 61:1600992. https://doi.org/10.1002/mnfr.201600992
Klein EY, Van Boeckel TP, Martinez EM et al (2018) Global increase and geographic convergence in antibiotic consumption between 2000 and 2015. Proc Natl Acad Sci U S A 115:E3463–E3470. https://doi.org/10.1073/pnas.1717295115
Kondratova AA, Kondratov RV (2012) The circadian clock and pathology of the ageing brain. Nat Rev Neurosci 13:325–335. https://doi.org/10.1038/nrn3208
Konjar S, Frising UC, Ferreira C et al (2018) Mitochondria maintain controlled activation state of epithelial-resident T lymphocytes. Sci Immunol 3:eaan2543. https://doi.org/10.1126/sciimmunol.aan2543
Kramer P, Bressan P (2015) Humans as superorganisms: how microbes, viruses, imprinted genes, and other selfish entities shape our behavior. Perspect Psychol Sci 10:464–481. https://doi.org/10.1177/1745691615583131
Kramer P, Bressan P (2018) Our (mother's) mitochondria and our mind. Perspect Psychol Sci 13:88–100. https://doi.org/10.1177/1745691617718356
Laje R, Agostino PV, Golombek DA (2018) The times of our lives: interaction among different biological periodicities. Front Integr Neurosci 12:10. https://doi.org/10.3389/fnint.2018.00010
Lane N (2005) Power, sex, suicide: mitochondria and the meaning of life. Oxford University Press, Oxford
Lane N (2011) The costs of breathing. Science 334:184–185. https://doi.org/10.1126/science.1214012
Lane N (2015) The vital question: why is life the way it is? Profile Books, London
Lemaître JF, Berger V, Bonenfant C et al (2015) Early-late life trade-offs and the evolution of ageing in the wild. Proc Biol Sci 282:20150209. https://doi.org/10.1098/rspb.2015.0209
Lewin N, Swanson EM, Williams BL et al (2017) Juvenile concentrations of IGF-1 predict life-history trade-offs in a wild mammal. Funct Ecol 31:894–902. https://doi.org/10.1111/1365-2435.12808
Linden DR (2014) Hydrogen sulfide signaling in the gastrointestinal tract. Antioxid Redox Signal 20:818–830. https://doi.org/10.1089/ars.2013.5312
Liu C, Chung M (2015) Genetics and epigenetics of circadian rhythms and their potential roles in neuropsychiatric disorders. Neurosci Bull 31:141–159. https://doi.org/10.1007/s12264-014-1495-3
Longo VD, Mattson MP (2014) Fasting: molecular mechanisms and clinical applications. Cell Metab 19:181–192. https://doi.org/10.1016/j.cmet.2013.12.008

López-Armada MJ, Riveiro-Naveira RR, Vaamonde-García C et al (2013) Mitochondrial dysfunction and the inflammatory response. Mitochondrion 13:106–118. https://doi.org/10.1016/j.mito.2013.01.003

López-Torres M, Barja G (2008) Lowered methionine ingestion as responsible for the decrease in rodent mitochondrial oxidative stress in protein and dietary restriction. Biochim Biophys Acta 1780:1337–1347. https://doi.org/10.1016/j.bbagen.2008.01.007

Martin WF, Neukirchen S, Zimorski V et al (2016) Energy for two: new archaeal lineages and the origin of mitochondria. Bioessays 38:850–856. https://doi.org/10.1002/bies.201600089

Martínez-Redondo D, Marcuello A, Casajús JA et al (2010) Human mitochondrial haplogroup H: the highest VO2max consumer – is it a paradox? Mitochondrion 10:102–107. https://doi.org/10.1016/j.mito.2009.11.005

Maruszak A, Adamczyk JG, Siewierski M et al (2014) Mitochondrial DNA variation is associated with elite athletic status in the Polish population. Scand J Med Sci Sports 24:311–318. https://doi.org/10.1111/sms.12012

Matheu A, Maraver A, Klatt P et al (2007) Delayed ageing through damage protection by the Arf/p53 pathway. Nature 448:375–379. https://doi.org/10.1038/nature05949

Merry TL, Ristow M (2016) Do antioxidant supplements interfere with skeletal muscle adaptation to exercise training? J Physiol 594:5135–5147. https://doi.org/10.1113/jp270654

Meyer JN, Leung MC, Rooney JP et al (2013) Mitochondria as a target of environmental toxicants. Toxicol Sci 134:1–17. https://doi.org/10.1093/toxsci/kft102

Miller VJ, Villamena FA, Volek JS (2018) Nutritional ketosis and mitohormesis: potential implications for mitochondrial function and human health. J Nutr Metab 2018:1–27. https://doi.org/10.1155/2018/5157645

Mostafalou S, Abdollahi M (2013) Pesticides and human chronic diseases: evidences, mechanisms, and perspectives. Toxicol Appl Pharmacol 268:157–177. https://doi.org/10.1016/j.taap.2013.01.025

Mottawea W, Chiang C-K, Mühlbauer M et al (2016) Altered intestinal microbiota–host mitochondria crosstalk in new onset Crohn's disease. Nat Commun 7:13419. https://doi.org/10.1038/ncomms13419

Myers JH (1978) Sex ratio adjustment under food stress: maximization of quality or numbers of offspring. Am Nat 112:381–388

Navara KJ (2014) Low gestational weight gain skews human sex ratios towards females. PLoS One 9:e114304. https://doi.org/10.1371/journal.pone.0114304

Newman JC, Verdin E (2014) Ketone bodies as signaling metabolites. Trends Endocrinol Metab 25:42–52. https://doi.org/10.1016/j.tem.2013.09.002

Omote H, Sambonmatsu N, Saito K et al (1999) The γ-subunit rotation and torque generation in F1-ATPase from wild-type or uncoupled mutant *Escherichia coli*. Proc Natl Acad Sci U S A 96:7780–7784

O'Neill JS, Reddy AB (2011) Circadian clocks in human red blood cells. Nature 469:498–503. https://doi.org/10.1038/nature09702

Pal VK, Bandyopadhyay P, Singh A (2018) Hydrogen sulfide in physiology and pathogenesis of bacteria and viruses. IUBMB Life 70:393–410. https://doi.org/10.1002/iub.1740

Pamplona R, Barja G (2006) Mitochondrial oxidative stress, aging and caloric restriction: the protein and methionine connection. Biochim Biophys Acta 1757:496–508. https://doi.org/10.1016/j.bbabio.2006.01.009

Penn DJ, Smith KR (2007) Differential fitness costs of reproduction between the sexes. Proc Natl Acad Sci U S A 104:553–558. https://doi.org/10.1073/pnas.0609301103

Peternelj TT, Coombes JS (2011) Antioxidant supplementation during exercise training: beneficial or detrimental? Sports Med 41:1043–1069. https://doi.org/10.2165/11594400-000000000-00000

Pitnick S, Karr TL (1998) Paternal products and by-products in Drosophila development. Proc Biol Sci 265:821–826. https://doi.org/10.1098/rspb.1998.0366

Powers SK, Jackson MJ (2008) Exercise-induced oxidative stress: cellular mechanisms and impact on muscle force production. Physiol Rev 88:1243–1276. https://doi.org/10.1152/physrev.00031.2007

Pyle A, Hudson G, Wilson IJ et al (2015) Extreme-depth re-sequencing of mitochondrial DNA finds no evidence of paternal transmission in humans. PLoS Genet 11:e1005040. https://doi.org/10.1371/journal.pgen.1005040

Radak Z, Chung HY, Koltai E et al (2008) Exercise, oxidative stress and hormesis. Ageing Res Rev 7:34–42. https://doi.org/10.1016/j.arr.2007.04.004

Radak Z, Zhao Z, Koltai E et al (2013) Oxygen consumption and usage during physical exercise: the balance between oxidative stress and ROS-dependent adaptive signaling. Antioxid Redox Signal 18:1208–1246. https://doi.org/10.1089/ars.2011.4498

Radak Z, Ishihara K, Tekus E et al (2017) Exercise, oxidants, and antioxidants change the shape of the bell-shaped hormesis curve. Redox Biol 12:285–290. https://doi.org/10.1016/j.redox.2017.02.015

Rahman J, Rahman S (2018) Mitochondrial medicine in the omics era. Lancet 391:2560–2574. https://doi.org/10.1016/S0140-6736(18)30727-X

Ramanathan C, Kathale ND, Liu D et al (2018) mTOR signaling regulates central and peripheral circadian clock function. PLoS Genet 14:e1007369. https://doi.org/10.1371/journal.pgen.1007369

Reddy AB, Rey G (2014) Metabolic and nontranscriptional circadian clocks: eukaryotes. Annu Rev Biochem 83:165–189. https://doi.org/10.1146/annurev-biochem-060713-035623

Redman LM, Smith SR, Burton JH et al (2018) Metabolic slowing and reduced oxidative damage with sustained caloric restriction support the rate of living and oxidative damage theories of aging. Cell Metab 27:805–815.e4. https://doi.org/10.1016/j.cmet.2018.02.019

Reiter RJ, Rosales-Corral S, Tan DX et al (2017) Melatonin as a mitochondria-targeted antioxidant: one of evolution's best ideas. Cell Mol Life Sci 74:3863–3881. https://doi.org/10.1007/s00018-017-2609-7

Rich P (2003) The cost of living. Nature 421:583. https://doi.org/10.1038/421583a

Ristow M, Zarse K, Oberbach A et al (2009) Antioxidants prevent health-promoting effects of physical exercise in humans. Proc Natl Acad Sci U S A 106:8665–8670. https://doi.org/10.1073/pnas.0903485106

Roenneberg T, Merrow M (2016) The circadian clock and human health. Curr Biol 26:R432–R443. https://doi.org/10.1016/j.cub.2016.04.011

Rogers GB, Keating DJ, Young RL et al (2016) From gut dysbiosis to altered brain function and mental illness: mechanisms and pathways. Mol Psychiatry 21:738–748. https://doi.org/10.1038/mp.2016.50

Rutter J, Reick M, McKnight SL (2002) Metabolism and the control of circadian rhythms. Annu Rev Biochem 71:307–331. https://doi.org/10.1146/annurev.biochem.71.090501.142857

Sato M, Sato K (2013) Maternal inheritance of mitochondrial DNA by diverse mechanisms to eliminate paternal mitochondrial DNA. Biochim Biophys Acta 1833:1979–1984. https://doi.org/10.1016/j.bbamcr.2013.03.010

Schulz TJ, Zarse K, Voigt A et al (2007) Glucose restriction extends *Caenorhabditis elegans* life span by inducing mitochondrial respiration and increasing oxidative stress. Cell Metab 6:280–293. https://doi.org/10.1016/j.cmet.2007.08.011

Sender R, Fuchs S, Milo R (2016) Revised estimates for the number of human and bacteria cells in the body. PLoS Biol 14:e1002533. https://doi.org/10.1371/journal.pbio.1002533

Sharpley MS, Marciniak C, Eckel-Mahan K et al (2012) Heteroplasmy of mouse mtDNA is genetically unstable and results in altered behavior and cognition. Cell 151:333–343. https://doi.org/10.1016/j.cell.2012.09.004

Song S (2012) Does famine influence sex ratio at birth? Evidence from the 1959–1961 great leap forward famine in China. Proc Biol Sci 279:2883–2890. https://doi.org/10.1098/rspb.2012.0320

Song S (2015) Privation, stress, and human sex ratio at birth. Early Hum Dev 91:823–827. https://doi.org/10.1016/j.earlhumdev.2015.10.009

Soong RK, Bachand GD, Neves HP et al (2000) Powering an inorganic nanodevice with a biomolecular motor. Science 290:1555–1558

Stojković B, Sayadi A, Đorđević M et al (2017) Divergent evolution of life span associated with mitochondrial DNA evolution. Evolution 71:160–166. https://doi.org/10.1111/evo.13102

Strong M (2004) Protein nanomachines. PLoS Biol 2:e73

Stuart JA, Page MM (2010) Plasma IGF-1 is negatively correlated with body mass in a comparison of 36 mammalian species. Mech Ageing Dev 131:591–598. https://doi.org/10.1016/j.mad.2010.08.005

Suh Y, Atzmon G, Cho MO et al (2008) Functionally significant insulin-like growth factor I receptor mutations in centenarians. Proc Natl Acad Sci U S A 105:3438–3442. https://doi.org/10.1073/pnas.0705467105

Sulak M, Fong L, Mika K et al (2016) TP53 copy number expansion is associated with the evolution of increased body size and an enhanced DNA damage response in elephants. Elife 5:e11994

Sutovsky P, Moreno RD, Ramalho-Santos J et al (1999) Ubiquitin tag for sperm mitochondria. Nature 402:371–372. https://doi.org/10.1038/46466

Swanson EM, Dantzer B (2014) Insulin-like growth factor-1 is associated with life-history variation across Mammalia. Proc R Soc Lond B Biol Sci 281:20132458. https://doi.org/10.1098/rspb.2013.2458

Tabatabaie V, Atzmon G, Rajpathak SN et al (2011) Exceptional longevity is associated with decreased reproduction. Aging 3:1202–1205. https://doi.org/10.18632/aging.100415

Templeman NM, Murphy CT (2018) Regulation of reproduction and longevity by nutrient-sensing pathways. J Cell Biol 217:93–106. https://doi.org/10.1083/jcb.201707168

Tian X, Seluanov A, Gorbunova V (2017) Molecular mechanisms determining lifespan in short- and long-lived species. Trends Endocrinol Metab 28:722–734. https://doi.org/10.1016/j.tem.2017.07.004

Trewin A, Berry B, Wojtovich A (2018) Exercise and mitochondrial dynamics: keeping in shape with ROS and AMPK. Antioxidants 7:7. https://doi.org/10.3390/antiox7010007

Trivers RL, Willard DE (1973) Natural selection of parental ability to vary the sex ratio of offspring. Science 179:90–92

Ueno H, Suzuki T, Kinosita K et al (2005) ATP-driven stepwise rotation of FoF1-ATP synthase. Proc Natl Acad Sci U S A 102:1333–1338. https://doi.org/10.1073/pnas.0407857102

Villanueva C, Kross RD (2012) Antioxidant-induced stress. Int J Mol Sci 13:2091–2109. https://doi.org/10.3390/ijms13022091

Vitale G, Brugts MP, Ogliari G et al (2012) Low circulating IGF-I bioactivity is associated with human longevity: findings in centenarians' offspring. Aging 4:580–589. https://doi.org/10.18632/aging.100484

Wallace DC (2005) A mitochondrial paradigm of metabolic and degenerative diseases, aging, and cancer: a dawn for evolutionary medicine. Annu Rev Genet 39:359–407. https://doi.org/10.1146/annurev.genet.39.110304.095751

Wanders D, Burk DH, Cortez CC et al (2015) UCP1 is an essential mediator of the effects of methionine restriction on energy balance but not insulin sensitivity. FASEB J 29:2603–2615. https://doi.org/10.1096/fj.14-270348

Wang X, Ryu D, Houtkooper RH et al (2015) Antibiotic use and abuse: a threat to mitochondria and chloroplasts with impact on research, health, and environment. Bioessays 37:1045–1053. https://doi.org/10.1002/bies.201500071

Yao CK, Muir JG, Gibson PR (2016) Review article: insights into colonic protein fermentation, its modulation and potential health implications. Aliment Pharmacol Ther 43:181–196. https://doi.org/10.1111/apt.13456

Zarse K, Schmeisser S, Groth M et al (2012) Impaired insulin/IGF1 signaling extends life span by promoting mitochondrial L-proline catabolism to induce a transient ROS signal. Cell Metab 15:451–465. https://doi.org/10.1016/j.cmet.2012.02.013

The manufacturer's authorised representative in the EU is Springer Nature Customer Service Centre GmbH, Europaplatz 3, 69115 Heidelberg, Germany. If you have any concerns regarding our products, please contact ProductSafety@springernature.com

Printed and bound by CPI Group (UK) Ltd, Croydon, CR0 4YY
22/09/2026
02220990-0009